THE PHILOSOPHY OF BIOLOGY

Philosophy

———

Editor

PROFESSOR S. KÖRNER

jur.Dr., Ph.D., F.B.A.

Professor of Philosophy

University of Bristol and Yale University

THE PHILOSOPHY OF
BIOLOGY

Michael Ruse

Associate Professor of Philosophy
in the University of Guelph, Ontario

HUTCHINSON UNIVERSITY LIBRARY
LONDON

HUTCHINSON & CO (*Publishers*) **LTD**
3 Fitzroy Square, London W1

London Melbourne Sydney Auckland
Wellington Johannesburg Cape Town
and agencies throughout the world

First published 1973

This book has been set in Times type,
printed in Great Britain on smooth wove paper
by The Camelot Press Ltd, London and Southampton,
and bound by Wm. Brendon, Tiptree, Essex

ISBN 0 09 115220 8 (cased)
0 09 115221 6 (paper)

CONTENTS

ACKNOWLEDGEMENTS

Many people helped me to produce this book and to them I would tender my thanks. First, there is Stephan Körner, who asked me to contribute a volume to this series and who encouraged me constantly through its writing. Secondly, there are the people (philosophers and biologists) who read earlier versions of parts or all of this work and who gave me detailed criticisms. Amongst these are Peter Alexander, A. J. Cain, Donald Colless, David Hull, Hugh Lehman, Alex Michalos, and Roger Smook. Finally, my thanks go to Kristina Casalini, Judith Martin, and Barbara Michalos, who typed drafts of this book and who protected the reader from my appalling spelling.

I am grateful to George Allen and Unwin Ltd. for permission to quote from *The Ascent of Life* by T. A. Goudge; and W. H. Freeman and Company for permission to quote from *The Principles of Numerical Taxonomy* by R. R. Sokal and P. H. A. Sneath.

I

INTRODUCTION

The author of a book on the philosophy of biology need offer no excuse for the subject he has chosen, since few areas of philosophy have been so neglected in the past fifty years. This is a great pity, for both philosophers and biologists have much to gain from a study of the no-man's land between them. As it is, philosophers tend to be almost totally ignorant of the many, recent, exciting advances in biology, and biologists tend to be hostile or indifferent to the work of modern philosophers. Consequently, philosophers build castles on non-existent scientific foundations, and biologists, almost inevitably, fight the battles that philosophers fought twenty or more years ago. In this book, I hope to show the rich vein of problems waiting to be mined by philosophers and biologists, and I shall feel that I shall have succeeded in my aim if I can infect but a few others with my enthusiasm for the subject.

There can be little doubt that modern biology dates from the publication in 1859 of Charles Darwin's *Origin of Species*—the work in which he argued that the organic world we see about us today is essentially the product of a slow gradual evolutionary process, brought about by something which he called 'natural selection'. Just about every area of biology today owes a great debt to Darwin, and for this reason it might be thought that the best way to start is by plunging straight into an exposition and analysis of the modern version of Darwin's evolutionary theory, best known as the 'synthetic' theory of evolution. It is part of my plan to give such an exposition and analysis of the synthetic theory; but, for reasons which will become apparent later, I shall first look at the legacy left by another nineteenth-century biological giant, Gregor Mendel. When I have discussed in Chapter 2 the Mendelian theory of inheritance (or, more precisely, the neo-Mendelian theory of inheritance), I shall be better able to consider the nature of natural selection and to evaluate its worth to the modern biologist.

However, before turning to Mendel, there is one matter which needs a brief discussion in this introduction. There are today no hard and fast 'schools' of thought about the philosophical nature of biology. Nevertheless, like King Charles's head, one question keeps cropping up in discussions, namely whether or not biology is a science like the sciences of physics and chemistry. Since, in this book, we shall be examining this question from various angles, it will perhaps be useful to give here a very brief outline of what I think is still the dominant philosophical position on the theories of physics and chemistry. This is the position commonly known as 'logical empiricism'. Obviously, no attempt at completeness can be made, and some of the points raised here will get more detailed discussion later in the book. (See also Braithwaite, 1953; Nagel, 1961; Hempel, 1966.) There are four points I want to raise about the kinds of results supplied by physical sciences.

(1) It is usually felt that in their work physical scientists refer to entities of two kinds. The first kind includes such things as molecules, electrons, wave functions and charges. The second kind includes such things as pendulums, prisms and planets. The first kind are sometimes called 'theoretical', 'hypothetical' or 'non-observable' entities; the second kind are called 'observable', 'really existent' or 'non-theoretical' entities. Just what kind of distinction should be made between the different kinds of entity we shall consider later. The main thing is that most people feel that some such distinction should be made.

(2) It is argued that the theories of physical science contain two kinds of statement. There are *a priori*, necessarily true statements—statements of logic and mathematics. Also there are true, universal, empirical statements. These latter statements, although they are not logically or mathematically necessary, are felt to be in some sense necessary. This empirical necessity is often called 'nomic' necessity, and the statements are called 'laws'.

(3) It is commonly supposed that the theories of physical science are essentially 'axiomatic' or 'hypothetico-deductive' systems. This means that one starts with a number of statements as premises, that is, as unproven within one's system, and from these all the other statements of one's system are derived. Newtonian mechanics is usually taken to be the paradigmatic example of such a scientific system—statements like Newton's laws of motion are used as premises, and statements like Kepler's laws are found to follow (in some sense). It is also usually the case that the premises of physical scientific systems refer mainly or exclusively to the kinds of entities

which I have labelled 'theoretical', and talk of 'non-theoretical' entities occurs only in (some of) the derived statements. Hence, a physical theory contains 'bridge' principles, enabling one to go from talk of one kind of entity to talk of the other kind.

(4) Closely connected with the claims about the axiomatic nature of the theories, it is felt that explanations in the physical sciences are of a particular kind. Without going into great detail here, it would seem that basically it is claimed that an explanation consists of a derivation of a statement about the thing being explained from other statements, at least one of which is a law. Explanations of this type are commonly called 'covering-law' explanations and, as is well known, it is frequently argued that such explanations can serve as the basis for predictions.

Anyone with the smallest acquaintance with the philosophy of science will know that all of these four claims, even as applied to physics, have many critics (e.g. Hanson, 1958; Achinstein, 1971). Since this book will not be a general, elementary text in the philosophy of science, distinguishable only in that it is adorned with a few biological examples, most criticisms must be ignored. However, let me declare here that, with considerable reservations to be noted, I think there is truth in the above claims as applied to the physical sciences. Moreover, let me also declare that I think these claims apply in large measure to the biological sciences. At least, I think they apply far more than many writers about biology have supposed, and I suspect that where the claims fail for biology, they often fail for physics also. Thus, I do not think that with respect to these claims a very hard line can be drawn between different kinds of science. But, in any case, the reader must judge these matters for himself. And so that this may be done, let us now turn to biology.

2

MENDELIAN GENETICS

Mendel's own work, as is well known, went practically unnoticed for thirty years. However, after its rediscovery at the beginning of this century, a theory of heredity based on his ideas was developed in great depth and at a rapid speed. I shall call this theory 'Mendelian genetics' in order to distinguish it from the very modern 'molecular genetics', the subject of Chapter 10. I shall begin here by looking at the Mendelian unit of inheritance, the 'gene', and then I shall go on to consider a number of philosophical problems which arise out of my discussion. I must point out that initially, for reasons which will become apparent shortly, I shall be considering genetics solely from the viewpoint of a theory designed to explain the transmission of heritable characteristics. Other evidence pertinent to the theory will be introduced later. I must also point out that at this stage I shall ignore, in a somewhat cavalier manner, organisms which provide difficulties for Mendelian genetics. Examples of such organisms, together with a discussion of their significance, will also come later. (George, 1964, is a good introduction to both Mendelian and molecular genetics.)

2.1 *The Mendelian gene*

There are many sides to the gene—the central concept of the theory we are here considering—but, by definition, the gene seems to be the *unit of function*. By this is meant that, in a way, it is genes which are the ultimate causes of all heritable organic characteristics. A man has brown eyes—it is a function of his genes. A man has blue eyes—it is a function of his genes. A man has brown hair, black hair, straight hair, curly hair, no hair—it is all the result of the different genes which he carries. Nevertheless, it is important to notice, right from the beginning, that when the genes are defined as the *ultimate* causes of organic characteristics, it is not being claimed that the genes are the *sole* causes of such characteristics. They are, as it were, the organism's contribution to its own development. The environment has an equally vital role in development, and indeed a change in the environment can have just as drastic an effect on an organism's finished form as can a change in gene. For example, genes control height; but, as is well known, malnutrition during a child's early years can have just as stunting an effect on growth as can a change in gene.

It is supposed that in sexual organisms the genes are fairly evenly divided into two sets, and that each member of one set has one and only one mate in the other set. These mates are said to occupy the same 'locus'. It is also supposed that a particular instance of a gene can be repeated—in this respect genes are significantly similar to the small particles of physics, for it is supposed that genes can be absolutely identical to each other. They are not merely 'more or less alike'. The genes at a particular locus might be the same, but possibly they are different. The members of a set of different genes, all of which can occupy the same locus, are called 'alleles'. (Obviously, at most only two members of such a set can be at the same locus in any particular organism.) If an organism has identical genes at a particular locus, it is said to be 'homozygous' with respect to that locus. If the genes are not identical, then the organism is said to be 'heterozygous'. Sometimes, the heterozygote looks like the homozygote for one of the genes at the locus. In this case, the gene whose effects are shown in the heterozygote is said to be 'dominant' over its mate, which latter is said to be 'recessive'. It should be noted that dominance and recessiveness are not absolute terms. A gene could be dominant over a second gene, but recessive to a third. Also let us note that some genes affect more than one characteristic. These are known as 'pleiotropic' genes. Some other genes form sets jointly affecting a single characteristic, in such a way that the effects of individual genes are not separately distinguishable. These are 'polygenes'.

Turning now to a somewhat different aspect of the gene, it is argued that it is the gene which is the link between one generation and the next. In this sense then, the gene is the *unit of inheritance*. Each parent contributes to the offspring, and the transmission is governed by what are known as 'Mendel's laws'. These are as follows:

Mendel's first law (also known as the 'law of segregation')
For each sexual individual, each parent contributes one and only one of the genes at every locus. These genes come from the corresponding

loci in the parents, and the chance of any parental gene being transmitted is the same as the chance of the other gene at the same parental locus.

Mendel's second law (also known as the 'law of independent assortment')

The chances of an offspring receiving a particular gene from a particular parent are independent of the offspring's chances of receiving any other gene (at a different locus) from that parent (i.e. genes at each locus segregate independently of genes at other loci).

Two points are particularly worth noticing about these laws. First, it is assumed that the units of inheritance are passed on entire and uncontaminated from one generation to the next—the genes from the parents do not 'blend' in the individual at any point during its lifetime. In this sense, Mendelian genetics, unlike most of its predecessors, is 'particulate'. Secondly, neither of the laws as stated above is strictly true. Later in the chapter we shall look at some of the exceptions to the laws and consider the implications of the exceptions, in particular whether real laws can have exceptions! Here, let us take note only of a major revision which has to be made to Mendel's second law, for it is found that some genes do not segregate independently, and that consequently, the chances of a particular gene being transmitted to an offspring is indeed a function of which other genes are being transmitted. Geneticists account for these exceptions to Mendel's second law with the following hypothesis. It is supposed that the genes are on lines, that these lines come in pairs, that the genes at a locus are always on different members of a pair, and that at some point before reproduction the lines are matched up, with each gene facing its mate. It is then argued that genes from different pairs of lines segregate independently (as the second law states), but that genes on a particular line are transmitted to offspring together, unless a phenomenon known as 'crossing-over' occurs. At certain points, a pair of lines might break, and the free ends might join with the ends of the opposing pair. Thus, the genes from one part of a line might be transmitted with the genes from the other part of the paired line. It is also hypothesized that crossing-over can occur more than once, and hence, genes at opposite ends of a line will segregate nearly independently, whereas genes close together will nearly always segregate together. A question now arises. What if certain characteristics are always transmitted together? Geneticists assume that here we have a case of pleiotropism, that is, that they are caused by the same gene. We have, therefore, another aspect of the gene. It is the *smallest unit of crossing-over*, that is, it cannot be divided by crossing-over.[1]

Finally in this brief exposition of genetics, let us look at the question of the stability of the gene. It is argued that a gene is normally very stable, and that it, or a copy, can remain unchanged for many generations. However, it is believed that sometimes a gene does change (into a different kind of gene), and that hence, organisms carry new heritable characteristics. This change, known as 'mutation', gives yet one more characterization of the gene. It is *that which mutates*. From the point of view of genetics, it is supposed that mutation is instantaneous, and it is of the essence of the theory that the change is 'random'. By this is meant, not that the change is uncaused—indeed, several causes of mutation are known, and like causes are assumed for all other cases—but rather that a change in a gene is never a function of the particular needs of an organism. New heritable characteristics do not appear 'to order'. When mutation yields a new heritable characteristic, the organism must make of it what it can, for its origin was unrelated to its possessor's particular environmental predicament.

We have now before us but the barest skeleton of Mendelian genetics; however, already we are in a position to ask some questions of far-reaching philosophical importance. In particular, we can ask whether this biological theory of heredity seems in any sense to correspond to the logical empiricist conception of a physical theory, sketched briefly in Chapter 1. To answer this question, I shall take in turn the four logical empiricist claims which I picked out for special attention, considering the first two in this chapter, and the other two in the next. I turn immediately therefore to my first major philosophical problem, namely whether or not Mendelian genetics refers to entities falling into two different classes.

2.2 What kinds of entities is Mendelian genetics about?

We saw in the last chapter that it is commonly argued that the entities referred to by physical theories seem to be of two kinds. On the one hand, we have things like molecules, electrons and charges; on the other hand, we have things like prisms, pendulums and planets. Now, *prima facie*, it would seem that in Mendelian genetics we find reference to entities of each of these kinds. Of the same kind as something like the molecule, we have the gene (and, more generally, the whole collection of an organism's genes, known as the 'genotype'.) Of the same kind as something like a pendulum, we

---

[1] Actually the full story of crossing-over is rather more complex than that presented here. In particular before crossing-over occurs there is a doubling of the genetic material. Hence, four lines in two pairs are involved in crossing-over. We need not pursue this complicating factor here.

have an organism's physical characteristics (known collectively as the 'phenotype'). Is our first intuition right in distinguishing between genotype and phenotype in this kind of way?

I shall suggest that basically this first intuition is correct; but I shall also suggest that the full story is rather complex, for, as several writers have pointed out recently, there seems to be no one absolute way of drawing a distinction between the entities of science—certainly, there is no single way which encompasses all of our intuitions. Rather, there are several ways of distinguishing between the entities of science, and whilst they are all in their own way informative, despite a considerable amount of overlap, each division has boundaries peculiar to itself, and indeed, the methods of division themselves are not entirely unambiguous. Before we turn to the genotype–phenotype dichotomy, let us examine this general point in more detail by considering three of the more popular criteria of demarcation for scientific entities. (See also Spector, 1966; Achinstein, 1968.)

Perhaps the most obvious way of making explicit one's intuition that the entities of science fall into two groups is in terms of one's ability to observe or see the various entities. Thus the members of the first set of entities (i.e. the molecule-including-set) are *unobservable*, whereas the members of the second set (i.e. the pendulum-including-set) are *observable*. Unfortunately, a little thought soon shows that this criterion of demarcation is not quite as straightforward as might appear at first sight. Consider, for example, a number of scientists taking note (in some fashion) of a lunar phenomenon (say an eclipse). Scientists A, B and C journey to South America, where A stands in the open and looks at the eclipse with his own eyes as does B (but who wears glasses), whereas C goes into the observatory and looks at the eclipse through a light telescope (involving lenses and perhaps reflectors). Scientist D stays at home and watches the eclipse on a simultaneous television transmission, E looks only at films brought back after the eclipse, F looks only at still photographs and G reads only reports. It does seem that scientists A, B and C can properly say that in some fashion they have observed the eclipse, and I think that most people would be prepared to allow that D was observing, albeit in a rather qualified sense—certainly a lot of us claim to have seen the first men land on the moon, although we were hardly there. I suppose that if one really presses the point, one can say even of E and F that for them the eclipse was, in a rather weak way, an 'observable' or 'observed' phenomenon, although one is clearly stretching the meaning of the term here. But if we allow at least some of these different senses of 'observe'—and we clearly must allow at least more than A's kind of observing, for it is indubitable that, say, C observed the eclipse in a very real way that G did not —then it obviously follows that one's distinction of entities into observable and unobservable is going to be very closely tied to the particular sense of 'observation' that one is using. In my example (which certainly does not explore every kind of observation) the set of observable entities for A-type observation will be less than but included in the set of observable entities for B-type observation, and so on, roughly speaking, right on up to F-type observation.

It might be objected that my example misses the point in that unobservable entities are supposed to be entities which are unobservable *in principle*, whereas my discussion has centred on a phenomenon, a lunar eclipse, which is or was observable in principle by any of the scientists A–G. But leaving aside the question of whether a short-sighted man is capable *in principle* of seeing distant objects (particularly phenomena which have to be viewed from a distance, like eclipses and rainbows) or whether one can in principle directly observe phenomena which are past, even if we grant this point, it does not affect my discussion very much. Consider scientist C_1 (analogous to C) who is looking at a very small object through a light microscope. He is surely observing this object in a way which is far more real and direct than a colleague G_1 who reads about C_1's findings in a journal. After all, any observation however direct is going to entail light rays passing through a transparent medium— it would be strange, not to say arbitrary, if the fact that this medium was glass as well as air were immediately to take the object being studied right out of our world of direct experience. But if these points be granted, then some objects which are in principle unobservable when 'observable' means 'observable with the naked eye' become observable when 'observable' is understood in a rather more generous sense.

What I would therefore suggest is that the moral of all of this seems to be, not that one cannot talk about observable or unobservable entities, but that one must specify in advance just what one means by 'observable' and 'unobservable' in a particular context. Thus, for example, one might want to talk about a molecule being unobservable, if one by 'observable' means 'visible with the naked eye' or even if one grants the aid of light microscopes. On the other hand, if one's notion of 'observable' extends to the use of electron microscopes, photographic techniques, and so on, then, in a somewhat pickwickian sense, one might speak of certain macro-molecules as being 'observable', at least, in a way that smaller entities are not. Of course, this is not to deny that many things, like electrons, are going to come up as unobservable under just about every way of drawing the distinction between observables and unobservables and the converse holds of other things like pendulums. For this reason, it is important not to over-exaggerate the border-line cases. But the different possible ways of drawing the observable–unobservable distinction do point to the fact that there are a fair number of intermediate entities which are liable to fall on different sides of the line, depending on just how one draws the line.

Because the observable–non-observable dichotomy seems rather loose, one might try an altogether different tack. Perhaps the division of entities could be done in terms of whether or not they are *hypothetical*. Thus, for example, we really know that chairs, tables, and pendulums exist; but for things like molecules, atoms, and electric fields, we have to form 'hypotheses'. Thus, their existence is hypothetical—inferred, but never really known. However, whilst this distinction of entities into hypothetical and non-hypothetical does certainly capture some part of our intuitive division of scientific entities, it is not obvious that it is adequate to bear the whole burden of the division. Consider something like the hydrogen atom, which presumably we would want to put into our hypothetical set. Would one, in fact, want to say that this is a 'hypothetical' entity? I think that most physicists would want to say that they are absolutely convinced that the hydrogen atom exists—it is certainly not hypothetical in the sense that we normally use the term, a sense which indicates some real doubt, as for example when a prosecuting counsel might say, 'well, let us now consider what this second hypothetical intruder is supposed to have done'. In an analogous manner, there are some things which some ways of drawing the division might consider hypothetical, but which we would feel should fall in the other set. Consider, for example, the possibility of big oil deposits in the North of Canada. Although many scientists think this to be a real possibility, such oil is, at this stage, 'hypothetical', even though one's observational–non-observational dichotomy such oil might fall into the observational camp.

Again, none of these points is to deny that there is some sense to the hypothetical–non-hypothetical division, or to deny that the division, to a fair extent, duplicates the observational–non-observational division. (More accurately, we should say that different ways of making the one division duplicate different ways of making the other division.) Probably the great majority of entities which we would in some sense call 'non-observational' are, or at least were, hypothetical in some manner, and the same goes for observational and non-hypothetical entities. For example, the chronon, a quantum of time, may be still hypothetical, and it seems to be unobservable. Atoms and molecules, although long suspected and probably now non-hypothetical, were once hypothetical. They also seem to fall mainly on the side of non-observability, although it is worth noting that probably their change to the side of non hypotheticality is, in small part, a function of their move in the direction of increased senses of observability. Conversely, chairs and tables are not usually very hypothetical, and similarly, they are more or less observable.

A third way of capturing our intuitive division of scientific entities, the final way we shall consider, is in terms of some sort of *theoretical–non-theoretical* dichotomy. There are several ways in which this dichotomy could be spelt out, in particular in terms of complexity or familiarity to a layman; but perhaps the most obvious basis of the division stems from the sense in which we take a theoretical entity to be one referred to by a term closely tied to a particular scientific theory, whereas a non-theoretical entity is not. In a way, this seems the most hopeful of all the methods of dividing scientific entities, since it seems fairly clear that something like 'kinetic energy' or 'electron' does not make too much sense, unless one knows something of the theory in which it occurs, whereas knowledge of something like 'chair' or 'table' does not require an extensive scientific background. Moreover, this method of division does parallel closely the other methods. Observable entities, for instance, can be seen directly, before one knows any theory—or makes any hypotheses. Unobservable or hypothetical entities, however, come only to us through theories. Nevertheless, it has been argued by several philosophers recently that there is no such thing as a scientific concept which is not in some sense linked to a particular scientific theory, that is, which is not in some sense 'theory-laden'. It is certainly argued that many concepts which refer to things which would fall in the observable and non-hypothetical camps are theory-laden. Thus, for example, Hanson says of the term 'crater' that to say of something that it is a crater 'is to infuse theoretical astronomy into one's observations . . . To speak of a concavity as a crater is to commit oneself as to its origin, to say that its creation was quick, violent, explosive . . .' (Hanson, 1958, 180). On the other hand, one can certainly see craters and there is no doubt as to their existence.[2]

All in all, we can see that one can draw several distinctions between scientific entities, and in so doing we can reveal some extremely interesting things about them. However, although we may

---

[2] In fairness, it must be pointed out that the claim that all observation is theory-laden is a highly controversial one, and many philosophers oppose it. I have myself argued elsewhere (Ruse, 1970a) that in the nineteenth century many of the points of dispute between Special Creationists and Evolutionists involved non-theory-laden observations.

have an intuitive feeling that scientific entities are of two different sorts, and although all of these divisions capture some of the reasons behind our intuition, it seems clear that there is no one absolute way to divide scientific entities, and that different ways of dividing will probably yield somewhat different divisions, despite fundamental similarities. This being so, let us now turn our attention back to the genotype–phenotype distinction, and, beginning with the observable–non-observable dichotomy, see what light the preceding discussion throws on our intuitions.

In the way in which I have presented Mendelian genetics in this chapter, no mention at all has been made of observation, and so one might think that the Mendelian gene is a totally non-observed (and unobservable) entity. However, obviously, this is a distorted view of the situation. Certainly, inasmuch as one restricts observation to the level of observation without any instruments, then the Mendelian gene is an unobservable entity. We cannot just look at organisms and see genes. However, if one extends the notion of observation to cover the use of microscopes (particularly light microscopes), then the situation alters. It becomes apparent that organisms are made up of cells, that these cells have very ordered parts, and that inside one part of the cells (the 'nuclei') one finds pairs of string-like entities (the 'chromosomes') of the kind supposed for the carriers of the genes. It is also found that at the time of the formation of the sex cells, the chromosomes undergo a process analogous to crossing-over (they form 'chiasmata'), that the sex-cells contain just one of each pair of chromosomes, and that each individual is the result of two such sex-cells, one from each parent, combining. Finally, although mutation of the gene cannot be observed, it can be seen that sometimes the chromosomes get disrupted and that this can lead to phenotypic change. Clearly, therefore, if one identifies the chromosomes with the carriers of Mendelian genes, one can observe much of what is claimed by Mendelian genetics, and one can answer some questions left dangling by my earlier exposition of Mendelian genetics. In particular, one can show that each organism does not have just one set of genes, but has the same set repeated in every cell (sex-cells excluded).

However, having granted this much, even with such an extended sense of 'observation', it would clearly be untrue just to conclude that the gene is an observable entity. For a start, much of our knowledge of the cell and of the chromosomes is obtained indirectly. It is rarely the crucial sex-cells which are observed, but others, more amenable to microscopic study. From these latter cells, the nature of the sex-cells is inferred. For instance, one of the most popular subjects of geneticists is the fruit-fly, *Drosophila*; but most of our very extensive knowledge about the chromosomes of the fruit-fly is obtained by study of the giant chromosomes in the salivary glands of the larvae. Secondly, even with our 'cytological' knowledge (i.e. knowledge of the cell), this is still a far way from being able to say we can observe the gene itself or that we can observe the many sides to the gene, some of which we listed earlier in the chapter. To see the carrier of the gene (the chromosome) is not necessarily to see the gene itself, for as is well known, and as will be discussed in some detail in Chapter 10, it is today believed that the gene consists of a long macromolecule of a substance called 'deoxyribonucleic acid' (DNA), and that strands of DNA are twisted round each other in a double helix. One certainly cannot see this with a light microscope. Admittedly, recently it has been claimed that one can 'see' the helix after a fashion—at least supposedly, one can photograph the shadow of the DNA molecule with the aid of an electron microscope, if first one coats the molecule with some heavy metal. However, obviously, even if one does claim that this is 'observation' of the gene, then this is 'observation' of an extremely technical kind—it is about the weakest sense of observation if all one has seen is something's photograph, let alone just a photograph of its shadow. Is a photograph of the Black Prince's armour's shadow an observation of the Black Prince, even if he is inside? Moreover, even observation of this kind does not, for example, show the gene as a unit of function.

It is clear, therefore, that if one allows that one can observe with the aid of instruments (and I think one should allow this, at least in some cases), there is some observational evidence for the Mendelian gene. (I mean here direct observational evidence—there is obviously lots of indirect observational evidence, like that given in breeding experiments.) However, even conceding this much, the gene is to a large extent an unobservable entity.[3]

Let us turn now to the question of the phenotype. Initially, one might think that there is little doubt but that the morphological and behavioural characteristics of organisms belong to the observable dimension of experience. Obviously, in a great measure this is true. One can observe the red belly of the sexually active male stickleback, just as one can observe his elaborate mating ritual. However, there

---

[3] Of course, if we identify the gene with the DNA molecule and consider the gene to be no more than this, then it is neither surprising nor interesting that the gene turns out to have the properties possessed by the small entities of physics. However, at this point I am not making the identification, but am considering the gene only as a biological phenomenon. My reference to the DNA molecule is intended merely to point to the fact that possibly there is some kind of observational evidence that the biological gene has a helical form—the physicochemical implications of this form have no relevance at this point, because we are still within the biological realm.

are at least a couple of items which show that the answer to the question about the nature of the phenotype is not entirely un-problematical. For a start, many organisms cannot be observed by the naked eye—microbes and viruses, for example. Some of these are so small that one can only properly 'see' them with the aid of an electron microscope. Thus, for example, even the phenotype of the phage T4 can be classified as observable only in the rather weak sense where one is prepared to allow that observation can involve the use of extremely sophisticated electronic magnifying instruments (the same, of course, goes for many of the phenotypic properties of larger organisms). For a somewhat different reason, most parts of the phenotypes of many normal-sized organisms are, in at least one important respect, unobservable for the biologist. By definition, the initial subject matter of the paleontologist is the remains of organisms long since dead. Now, it may indeed be the case that the skilled paleontologist can reconstruct the nature and habits of an organism from a few scattered fossilized bones; but it is perhaps a little odd to say that he 'observes', say, an instance of *Archaeopteryx*—he certainly does not observe in it the way that I observe instances of *Passer domesticus* (house-sparrows) from my breakfast table. It is perhaps even odder to say that the paleontologist observes organisms when he infers their existence from things like the remains of their tracks and their feces.[4]

All in all, it does seem that the observable–non-observable dichotomy, or rather, different ways of drawing the dichotomy capture some if not most of the difference between phenotype and genotype; but it is clearly not the case that phenotypes fall neatly one one side of all of the divisions, and genotypes on the other side. The sorts of things which seem to prevent a clear-cut division in the inorganic world seem to have similar effects in the organic world. With this conclusion, let us look briefly at the other two ways in which we might try to make the division between phenotype and genotype.

As far as the hypothetical–non-hypothetical dichotomy is concerned, there is no doubt that at one point, until about 1910, biologists considered genes to be hypothetical entities, and they refused to credit them with anything so vulgar as existence (Carlson, 1966). However, as the evidence piled up, and particularly as the cytological data poured in, there grew less and less doubt in anyone's mind that genes really do exist—in fact, I feel sure that today no Mendelian geneticist doubts the existence of genes any more than he does that of chairs and tables. Of course, this is not to say that

---

[4] Of course, this is not to deny that, with the aid of a time-machine, the paleontologist could directly observe his organisms; but the whole point is that normally we do not presuppose time-machines when we talk of 'observation'.

at any point a Mendelian geneticist would want (or have wanted) to claim that he has (or had) full knowledge of the gene in general, or of any gene in particular. For example, the exact nature of crossing-over has until recently been in doubt, and the exact causes of mutation have long been a subject of intense speculation. Hence, although the existence of the gene as such might not be very hypothetical, it was at one point, and some properties still attributed to it are, and the same goes for the existence of certain individual genes.

On the other side of the coin, by and large there does not seem to be too much hypothetical about the nature of phenotypes. One does not have much doubt about the existence of the stripes of the zebra. However, there are some exceptions to this rule. Many organic behavioural characteristics, for example, have long been (and, indeed, some still are) in doubt. Moreover, the whole existence of microbes was taken to be hypothetical (if not outrightly false) until the work of people like Pasteur, and a similar debate occurred at the end of the seventeenth century over whether or not fossils really are the remains of long dead organisms. Even today, paleontologists have to hypothesize about many of the characteristics of their subjects— particularly about perishable things like skin and behaviour. Consequently, although in the main the phenotype is not hypothetical, some phenotypes have been taken to be so and some parts of some phenotypes still are.

Finally, there is the theoretical–non-theoretical dichotomy. Here also, the dichotomy roughly distinguishes between the genotype and the phenotype, but not exactly. Obviously, the Mendelian gene is central to Mendelian genetics, and similarly, without some knowledge of the discipline one cannot claim to know much about the entity. Conversely, one does not seem to need much genetical knowledge to talk of a leg, arm or whole organism, like an instance of *Drosophila melanogaster*. However, even here, matters are not entirely clear-cut. We shall see in later chapters that a full understanding of organic characteristics might well require of one knowledge of some theory —in particular, talk of characters can presuppose a knowledge of the *function* that they serve, and this, as we shall see, commits one to a particular theory of evolution. We shall also see that to talk of a particular kind of organism (as, for example, when one says that only a limited number of genes separate *Drosophila pseudoobscura* from *D. miranda*) is, according to certain writers, to commit oneself to a particular kind of history (which is, in turn, inferrable only through a particular theory). No more need be said at this point, except that once again the dichotomy makes roughly the same division as the other dichotomies, but that such a division does not seem unambiguously clear-cut.

To bring to an end our discussion in this section, let us recap the main points. It would seem that although intuitively one can draw a division between entities referred to by physical theories, many questions arise when one tries to specify the exact basis of this division. Questions about the observability of entities, about the hypothetical nature of entities, and about the theory dependence of entities, all seem to be involved, and although divisions based on these separate considerations do roughly coincide, the precise groups one draws can depend on how exactly one interprets the considerations themselves. Moreover, much of the analysis as it applies to the entities of physics and chemistry seems directly applicable to the entities of Mendelian genetics. As we saw, the three criteria of demarcation do seem to distinguish genotype and phenotype. However, as we also saw, such a division is rather rough, and the full story is somewhat more complex. If one allows the use of instruments, then many aspects of the gene are observable, whereas even with instruments some phenotypes are unobservable in some senses, although this is not to deny that in many basic respects genotypes are unobservable whereas phenotypes are observable. The gene itself may not be hypothetical, but it was, and many aspects of it still are; conversely, generally speaking phenotypes are not and never were hypothetical, although some were, some still are, and the same holds of some phenotypic characteristics. Finally, the gene does seem to be strongly linked with theory, much more so than phenotypes; but perhaps even here, because of some phenotypic theory-ladenness, the full story is rather more complex.

Hence, it would seem that, on the basis of our discussion, Mendelian genes seem very similar to many of the entities of the physical sciences, in particular, to things like molecules. This holds both by virtue of the fact that genotypes seem to be on the same sides of the divisions just discussed as molecules, and by virtue of the fact that the divisions do not give unambiguous answers in all cases and the ambiguities have the same kinds of sources (e.g. ambiguities stemming from just what kind of instruments one will allow and still be prepared to call an entity 'observational'). Conversely, in this respect, phenotypes are similar to things like pendulums. With this conclusions let us now go on to the next aspect of Mendelian genetics to be discussed. This revolves around the second logical empiricist claim I mentioned, that concerning laws.

2.3   *Does Mendelian genetics contain laws?*

We have seen that Mendelian genetics certainly contains things which biologists call 'laws', namely Mendel's two laws. However, at this point we encounter our first critic of biology, for recently one influential philsopher of science has argued that the so-called 'laws' of biology (including those of Mendelian genetics) are really not laws at all. In his book *Philosophy and Scientific Realism,* Smart argues that if something is really a law (as opposed to something merely called a 'law'), then there are at least two conditions which it must satisfy. The first condition is one which we might call the condition of 'unrestricted universality', and it specifies that a law must apply truly throughout the universe and throughout time. The second condition is that a law must make no reference, explicit or implicit, to any particular place or thing (in particular, it must not refer to the Earth). Smart believes that the would-be laws of biology all fail on at least one of these two conditions. Normally, they make reference to the Earth, and if one removes this reference, then the chances are that they are not true elsewhere in the universe.

To illustrate his point, Smart invites the reader to consider (what he takes to be) a typical, non-analytic, general statement of biology, 'albinotic mice always breed true'. About this statement, supposedly considered a law by biologists, he writes: 'What are mice? They are a particular sort of terrestrial animal united by certain kinship relations. They are defined as mice by their place in the evolutionary tree . . . The word "mouse" therefore carries implicit reference to our particular planet, Earth' (Smart, 1963, 53). Hence, we have no real law.

The obvious counter to this criticism is to redefine 'mouse' without reference to the Earth (e.g. by a set of properties A_1, A_2, ... A_n); but this move entails, Smart believes, a violation of the condition of unrestricted universality.

No doubt we could find a set of properties such that, so far as terrestrial animals are concerned, all and only mice possessed them. The trouble is that now we have no reason to suppose our law to be true. The proposition that everything which possesses the properties A_1, A_2, ... A_n and which is albinotic also breeds true is very likely a false one . . . on some planet belonging to a remote star there may well be a species of animals with the properties A_1, A_2, ... A_n and of being albinotic but *without* the property of being true. (Smart, 1963, 54)

Hence, again we have no real law.

In a like manner, Mendel's laws do not escape Smart's critical attention, and these too he relegates to the limbo of non-nomic, universal generalizations. He writes about Mendel's law of segregation as follows:

Even terrestrial populations do not segregate quite in accordance with the Mendelian principle, for a multitude of reasons, of which the chief is the phenomenon of crossing over. Even if we tried to protect our law by adding clauses such as 'if there is no crossing over', we should be pretty sure to be caught out by some queer method of reproduction obtaining on other spheres. Of course, there may well be good reasons why life on other worlds must be expected to have a rather similar chemical constitution to life on ours. Perhaps in every case we may expect it to have begun with the creation of amino-acids and the combination of these into larger molecules. Nevertheless, it would be altogether too speculative to assert that things have always gone on in other planets as they have done here, and that, for example, the genetic codes are necessarily embodied in nucleic acid molecules as is the case here. Perhaps so, perhaps not. In any case, we are here talking at the biochemical level. (Smart, 1963, 56)

This charge that Smart is making is extremely serious, for it is perfectly obvious that, as Smart cheerfully admits, if there are no laws in biology, then any attempts to cast biology (in particular, genetics) into a nice, neat, formal, axiomatic system are going to look a little forced. The non-analytic propositions of biology will be far too 'flabby' for anything like that. We shall have to resign ourselves to the fact that biology is unalterably different from physics and chemistry, where 'different' in this context is a euphemism for 'second-rate'.

Fortunately, I think it can be proven that Smart is quite mistaken in all of the claims which are important to his case. In order to show this I shall, in this section, first consider what it is that makes us call something a 'law of nature', and then I shall show that there are no good reasons for denying this title to Mendel's laws. In Chapter 3, I shall consider Smart's subsidiary claim that biology (in particular, Mendelian genetics) cannot be truly axiomatic, and in showing that Smart is wrong, I shall give a further example of a biological law. Later in the book, I shall take up again the question of law, and shall discuss some much more controversial candidates which lie outside the area of the purely genetical.

What is a law of nature, or more particularly, what is it that distinguishes a law from other statements? We have seen that according to the logical empiricist account a law of nature must be a true, universal, non-analytic statement. But, of course, even if we accept this as a necessary condition of what it is to be a law, there are other true, universal, non-analytic statements which we would not want to call laws. Thus, for example, we would want to call Boyle's law ('For a fixed mass of gas at constant temperature, pressure × volume = constant') a 'law', just as we would want to call Snell's law ('whenever any ray of light is incident at the boundary separating two media, it is bent in such a manner that the ratio of the sine of the angle of incidence to the sine of the angle of refraction is always a constant quantity for those two media') a 'law'. However, even if we suppose that the statement (to give an example of Smart's) 'whenever one turns the left-hand knob of a radio one gets squeaking' is true, despite the fact that it is obviously also universal and non-analytic, we would not want to call it a 'law'. The reason is because, as was mentioned earlier, the laws seem in some sense necessary ('nomically' necessary), whereas the radio statement does not seem necessary—somehow, its truth is a 'matter of chance': things could quite well have been otherwise.

But let us ask ourselves now, wherein lies the source of the feeling of necessity that we have about laws? As Hume and Kant so forcefully pointed out, it is not something which we see, but is something which we have to supply ourselves. It is at this point, I think, that Smart's criterion of unrestricted universality comes in. Although there is no logical implication, we feel that were something to hold at any place and in any time, this holding would have to be something over and above a matter of mere contingency. (Whether or not we are right in feeling this is not my concern here. I am just trying to see if the 'laws' of biology are like the 'laws' of physics and chemistry, not whether the notion of 'law' in itself is a tenable concept.) In other words, what I am suggesting is that were something to satisfy the criterion of unrestricted universality, we would be prepared to call it a 'law' (let us leave to one side, for the moment, the other condition that Smart makes of laws).

Now, the obvious drawback to a mark of lawlikeness like the satisfaction of the criterion of unrestricted universality is that it is totally impractical. No one, as yet, can test putative laws on Andromeda, and no one ever will be able to test putative laws in the Devonian period. Thus, we must find some indirect way of convincing, say, a lover of radios, that his law candidate ('whenever one turns the left-hand knob of a radio one gets squeaking') is not a real law, whereas Boyle's law and Snell's law are. Clearly, what we must do is weaken the condition of unrestricted universality down to something within our own range of experience. At best, we can demand of something before we are prepared to call it a 'law', that it be found not to clash with any of our other beliefs (in particular, with theories that we accept), that it be found to hold in a wide range of different conditions, and hopefully, some of these conditions will be such that we never thought of the law holding under them, before we first conceived of the law. Because the law is found to be true of many different types of circumstances, some of which were not 'built-in' before we started, we think it will be true of circumstances throughout time and space, many of which will be quite beyond our imagination. The whole point is that Boyle's law and Snell's law have been found to satisfy such a wide range of situations (some of which were unknown before they were formulated)—the radio 'law' does not even satisfy this very limited version of the condition of un-restricted universality. Hence, I would suggest that it is on such evidence that our feelings of nomic necessity rest.[5]

Three additional points need briefly to be made before we can return to biology. In the first place, even though the lover of radio might finally relinquish his backing for his own 'law', he might with justice point to the fact that many of the things that the physicists would call 'laws' fail to satisfy even the limited condition I have given above. Consider, for example, the following statement suggested by Hempel: 'On any celestial body that has the same radius as the earth but twice its mass, free fall from rest conforms to the formula $s = 32t^2$' (Hempel, 1966, 57). There may well be no such celestial body; but probably physicists would still want to call the statement a 'law'. The answer to this problem is, I think, that physical scientists are happy to call something a 'law', even if the limited condition of unrestricted universality is satisfied only indirectly. If a statement is part of an axiomatic body of scientific theory or follows from such a theory (as Hempel's statement follows from Newtonian mechanics), then physical scientists are prepared to concede that the evidence in favour of the whole (which might be very great, as in the case of Newtonian mechanics) can count in favour of the part (which might have very little evidence, as in the case of Hempel's statement).

Secondly, one must mention the fact that although most accounts of what it is to be a law make part of the necessary condition of being a law that of being true, this is not strictly the case for nearly any law (i.e. a statement which is given the name 'law'). Neither Boyle's law nor Snell's law hold exactly for anything, and Boyle's law breaks down drastically at high temperatures and pressures, as does Snell's law for certain media, like Iceland Spar. Consequently, we must recognise that even the laws of physics are often only approximately true and they carry caveats excluding certain substances and so on. Moreover, there is frequently a kind of 'evolution' of laws as physicists try to diminish the degree of approximation that a law entails. (Think of the evolution of Boyle's law to van der Waal's equation and beyond.) Obviously, there is something of a tension in claiming on the one hand that laws are necessary, and on the other hand admitting that they are not true. Perhaps the best thing is to think of laws as being true within a certain range and to a certain approximation. But we must recognize these limitations, even in physics—most accounts ignore them (see Scriven, 1961).

[5] At this point I am talking of our *evidence* for laws. I am not suggesting that laws are *discovered* simply by looking at a whole lot of different examples.

Finally, there is the question of Smart's other condition for lawlikeness, namely that a law should not refer to any particular time or place. Now, like it or not, Smart must recognize that there are certain things which physicists call 'laws' which do refer to particular things—Kepler's laws are the paradigmatic example. Perhaps the best compromise is to distinguish, as do Hempel and Oppenheim, between 'fundamental' and 'derivative' laws (Hempel and Oppenheim, 1948). Fundamental laws are those which satisfy Smart's second criterion (as do, for example, Newton's laws), and derivative laws make reference to a place or time and are derived from fundamental laws together with other assumptions (as are Kepler's laws). It should be noted that this is a compromise, for Kepler's laws were considered 'laws', before they were shown to be derivable (after a fashion) from Newton's laws.

This brief discussion about the nature of laws concluded, let us now turn our attention back to biology. In view of the ground that has been covered, does Smart's claim that Mendel's laws are not real laws hold water? I would suggest, quite emphatically, that it does not. For a start, there can be a few statements of science which have been found to satisfy the limited condition of unrestricted universality more fully. Since Mendel first proposed his laws, they have been found to hold for a range of organisms from elephants to cod-fish, from sea-weed to oak-trees. Moreover, nearly all of this evidence came after Mendel discovered the laws—the circumstances were not built into his formulation. Mendel worked with a very limited range of organisms—mainly pea plants. And indeed, Mendel's published results are probably too good to be true, and hence he may well have discovered the laws before he did most of his experiments on the pea plants (Fisher, 1936; Wright, 1966).

Looking at the matter from the other side, let us take the specific charges Smart levels against Mendel's laws. He argues that the law of segregation breaks down because of the phenomenon of crossing-over, and that, for this reason, it cannot be a genuine law. Even if we repair the 'law' so that it is no longer false, then we will probably get caught out elsewhere. In any case, too much by way of renovation will probably reduce us to talk at the biochemical level, or Smart later adds, to the conversion of our laws into tautologies. 'If we try to produce laws in the strict sense which describe evolutionary processes anywhere and anywhen it would seem that we can do so only by turning our propositions into mere tautologies' (Smart, 1963, 59).

In reply to Smart, it is worth noting first of all that, as we have seen, it was the law of independent assortment which stood in need of revision and that, in any case, the problem was linkage, not crossing-over. Secondly, it seems unfair to condemn this 'law' because it has stood in need of revision, when laws like Boyle's law and Snell's law have stood equally in need of revision. It is not as if what went wrong with the 'law' was something for which we have no reason and that such a breakdown could easily happen at any-time again. We know now that the exceptions are due to genes being linked on the same chromosome, and it is worth mentioning that it is also now known that these genes obey extremely strict laws of their own. In fact, because the laws governing genes on the same chromosome are obeyed so strictly, geneticists have been able to 'map' the order of genes on chromosomes with great accuracy.[6]

Thirdly against Smart, whilst it must be admitted that there are other exceptions to Mendel's laws, most particularly non-chromosomal genes which do not obey the first law, although such exceptions obviously interest biologists greatly, these exceptions prove no more than one finds in similar cases for physical and chemical laws. (Mendel's laws are now usually stated in a form specifically restricting them to chromosomal genes.) Fourthly, Smart cannot avoid calling Mendel's laws 'biological laws' by pretending that they are now biochemical. Whatever may be the current relationship between Mendelian genetics and the physical sciences, there is nothing physico-chemical about the Mendelian gene. As I pointed out earlier, no knowledge of physics or chemistry was presupposed by my exposition of Mendelian genetics. Indeed, the excitement which scientists presently feel at the possibility of making genetics as conceived and presented part of the physical sciences stems precisely from the fact that Mendelian genetics is *not* part of the physical sciences. Fifthly and finally, whatever else they may be, Mendel's laws are not tautologies. Logically, it is possible for one parent always to supply three-quarters of the offspring's genes, and for no genes to segregate independently.

All in all, there seems little reason to agree with Smart that biology has no laws, and that Mendel's laws, in particular, are not lawlike (where this is understood in the sense of 'being laws like those of the physical sciences'). Moreover, using the terminology which we adopted earlier, it is worth noting that Mendel's laws are 'fundamental', that is, they make no reference to any particular

---

[6] In fairness, it must be admitted that through a refinement of the notion of crossing-over, the clauses dealing with the exceptions to Mendel's second law have been subject to the kind of evolution I mentioned in the context of the laws of physics. But this refinement, to be discussed in a later chapter, does seem to be no more than the sort one finds in physics, and hence it is no reason for a sharp division between physics and biology. In any case, the revision does not affect Mendel's second law, when it is understood as not applying to genes on the same chromosome.

place or time. In this sense, therefore, there is nothing peculiar about biology.[7]

Before concluding this stage of the discussion one point is worth noting. It must be allowed that Mendel's laws do have one feature not shared by every law of the physical sciences—unlike something like Boyle's law, they are *statistical* laws. They tell us about the probabilities of certain things happening (e.g. 'a gene has a 50 per cent chance of being transmitted'), rather than about their inevitability. Statistical laws occur in the physical sciences (e.g. 'the half-life of radon is 3·82 days'), so the existence of biological statistical laws do not in themselves separate off the biological sciences; but we shall see laws of this kind recurring in biological discussions throughout the book, and I think it would not be unfair to say that we find in biology a much greater reliance on statistical laws than we do in many parts of the physical sciences. Hence, I would make the concession (one certainly not forced on me by Smart) that, although the discussion has not yet revealed any basic differences between the physical and biological sciences, frequently statistics plays a greater role in biology than it does in most parts of physics and chemistry. But this, I would emphasize, seems to be a difference of quantity rather than quality. It is not a firm criterion of demarcation of the kind that Smart seeks.

[7] Smart's example about mice seems not to affect this conclusion, once one adopts the distinction between fundamental and derived laws. The example follows from Mendel's first law, together with certain assumptions about albinic genes, although I suspect that these assumptions might have to be so qualified that biologists might feel uneasy about calling the final product a 'law'—rather than, as Smart would have it, that they would call it a 'law' when it really is not one.

3

POPULATION GENETICS

Up to now our concern has been basically with the genetics of the individual. In this chapter, as we look at the final two claims I picked out from the logical empiricist thesis—first the question of the axiomatic nature of Mendelian genetics and secondly the question of explanation—our concern will start to veer away from the genetics of the individual towards the genetics of the group. There is of course no hard and fast distinction between these two kinds of genetics, nor do biologists attempt to draw one. However, customarily they refer to studies which are essentially concerned with group-heredity, as opposed to the study of the individual, as 'population genetics', where it is understood that this is a part of the whole field of Mendelian genetics. It will be this particular kind of genetics which is the subject of this chapter, and I shall take up where I have just left off and shall consider Smart's subsidiary claim, namely that no part of biology can properly have an axiomatic form. (Good source-books on population genetics are Li, 1955; Falconer, 1961.)

3.1 *Is Mendelian genetics a genuine axiomatic system?*

Backed by his attack on the notion of biological law, Smart writes:

Writers who have tried to axiomatize biological . . . theories seem to me to be barking up the same gum tree as would a man who tried to produce the first, second, and third laws of electronics, or of bridge building. We are not puzzled that there are no laws of electronics or of bridge building, though we recognize that the electronic engineer or bridge designer must use laws, namely laws of physics. The writers who have tried to axiomatize biology . . . have wrongly thought of biology . . . as a science of much the same logical character as physics, just as chemistry is. (Smart, 1963, 52)

Now since the sole basis for this claim of Smart is that there are no laws in biology, and since we have just shown at the end of the last chapter that there are indeed laws in biology, the rug has been pulled out from underneath him, somewhat. But of course, destroying Smart's argument is not in itself a proof of the fact that the part of biology being discussed has an axiomatic form, and so what I shall do now is give a simple example as evidence that not only *can* genetics be axiomatized, but that it is *in fact* axiomatized (in a perfectly unobjectionable manner). Incidental to this main task, I shall supply more evidence of the fact that biology contains (fundamental) laws, by supplying a statement which in this instance is considered a 'law' by virtue of its place within a scientific system. And also, I shall show that genetics contains bridge principles (between the genotype and phenotype), akin to those found in the physical sciences.

To say that something is axiomatized is to say that we start with some statements as premises (in science, these statements are laws together with logical and mathematical truths), and from these we derive other statements. The derivation I want to consider is that of a very important law of biology known as the Hardy–Weinberg law (H–W law, for short). The derivation is from Mendel's first law; from other biological assumptions, for example that male–female crosses are the same as female–male crosses (not always true, but within the limits set earlier); and from certain basic mathematical assumptions. In order to understand the H–W law we must suppose that we have a random-mating ('panmictic') population of sexual organisms. We must also suppose that the population is large enough to behave for all intents and purposes as if it were infinitely large, and for simplicity, let us confine ourselves to one gene locus and suppose that at this locus there can be either of two alleles, A_1 and A_2. This means that we can have individuals A_1A_1, A_1A_2, and A_2A_2 (the two homozygotes and the heterozygote). Also, let us suppose that at some point in the history of the population, the ratio of A_1 genes to A_2 is $p : q$. The H–W law then states that, supposing there are no outside influences and no mutations to or from the alleles involved, the gene ratio will stay at $p : q$, and, for all generations after the first, the genotypes will be divided as follows:

$$p^2 A_1A_1 : 2pq\ A_1A_2 : q^2\ A_2A_2$$

(i.e. in a representative sample of n organisms of the population, np^2 will have genes A_1A_1, $n2pq$ genes A_1A_2, and nq^2 genes A_2A_2). Since we are assuming A_1 and A_2 are the only genes at this locus, $p + q = 1$, and so the law can also be stated as:

$$p^2\ A_1A_1 : 2p(1-p)A_1A_2 : (1-p)^2A_2A_2$$

B

To see how this law is derived, let us assume that we have a population satisfying the antecedent conditions of the H–W law, and let us also assume that initial distribution of genes amongst the organisms is:

$$P\ A_1A_1 : H\ A_1A_2 : Q\ A_2A_2$$

(where $P + H + Q = 1$)
This means that we have:

	Genes		Organisms		
	A_1	A_2	A_1A_1	A_1A_2	A_2A_2
Frequencies	p	q	P	H	Q

Since an organism could mate in nine different ways, and since mating is random, we can draw up a table showing frequencies of types of mating as follows:

	Frequency of male organism		
	A_1A_1	A_1A_2	A_2A_2
	P	H	Q
Frequency of female organism			
A_1A_1 P	P^2	HP	QP
A_1A_2 H	PH	H^2	QH
A_2A_2 Q	PQ	HQ	Q^2

Now, if we assume Mendel's law of segregation, we can calculate the proportions of different offspring these matings will yield.
For example:

(1) If A_1A_1 breeds with A_1A_1 (symbolically $A_1A_1 \times A_1A_1$), the offspring must be A_1A_1.
(2) If $A_1A_1 \times A_1A_2$, then the offspring are $\frac{1}{2} A_1A_1$ and $\frac{1}{2} A_1A_2$.
(3) $A_1A_2 \times A_1A_2$ yields $\frac{1}{4} A_1A_1$, $\frac{1}{4} A_1A_2$, $\frac{1}{4} A_2A_1$, $\frac{1}{4} A_2A_2$.
(4) $A_1A_2 \times A_1A_1$ yields $\frac{1}{2} A_1A_2$ and $\frac{1}{2} A_1A_1$.

Similarly in the other five cases. But, in this context, a male with one genotype breeding with a female with another genotype is equivalent to a female with the first genotype breeding with a male with the second genotype, and obviously (although perhaps not analytically) an A_1A_2 genotype is equivalent to an A_2A_1 genotype. Thus we can draw up the following table showing the distribution of the genes in the offspring:

Mating		Frequency of offspring		
Type	Frequency	A_1A_1	A_1A_2	A_2A_2
$A_1A_1 \times A_1A_1$	P^2	P^2	—	—
$A_1A_1 \times A_1A_2$	$2PH$	PH	PH	—
$A_1A_1 \times A_2A_2$	$2PQ$	—	$2PQ$	—
$A_1A_2 \times A_1A_2$	H^2	$\frac{1}{4}H^2$	$\frac{1}{2}H^2$	$\frac{1}{4}H^2$
$A_1A_2 \times A_2A_2$	$2HQ$	—	HQ	HQ
$A_2A_2 \times A_2A_2$	Q^2	—	—	Q^2
SUMS		$(P + \frac{1}{2}H)^2$	$2(P + \frac{1}{2}H)(Q + \frac{1}{2}H)$	$(Q + \frac{1}{2}H)^2$

Now, the relationship between p and q, and P, H, and Q is easy to find. An organism with an A_1A_1 genotype has two A_1 genes, an A_1A_2 has one A_1 gene, and an A_2A_2 has none. Hence, given any representative N members of the population, they must have $(2P + H). N. A_1$ genes, and $(H + 2Q). N. A_2$ genes. Consequently, since they have $2N$ genes in all, the proportion of A_1 genes is $[(2P + H) N]/2N$, and the proportion of A_2 genes is $[(H + 2Q) N]/2N$. Thus, $p = P + \frac{1}{2}H$ and $q = \frac{1}{2}H + Q$. Hence, the distribution of the offspring is

$$(P + \tfrac{1}{2}H)^2 A_1A_1 : 2(P + \tfrac{1}{2}H)(Q + \tfrac{1}{2}H) A_1A_2 : (Q + \tfrac{1}{2}H)^2 A_2A_2$$

or, $p^2 A_1A_1 : 2pq A_1A_2 : q^2 A_2A_2$.

Having now obtained the ratio stated in the H–W law, in order to see that the gene ratio stays the same, suppose that the new ratio is $p_1 : q_1$. We have $p_1 \equiv p^2 + pq$, and $q_1 \equiv pq + q^2$, and hence $p_1 : q_1 = p : q$. Therefore, the H–W law is seen to hold for one generation, and obviously it will hold for all future generations. And, with the completion of this derivation, I submit both that we can now clearly see that at least parts of Mendelian genetics are axiomatized, and that some of the laws of genetics, in particular the H–W law, are to be considered lawlike as much as by virtue of their place within an axiomatic system as by virtue of any direct evidence which might exist in their favour.

Finally in this section, let us note that geneticists do seem to rely on bridge principles to take them from the level of the genotype to the level of the phenotype (and, where the phenotype is not directly observable, to the level of observation), although admittedly, they often do not state these principles very explicitly. An example of the use of such principles occurs in a study which was made on blood groups of a sample of 1,279 English people (Race and Sanger, 1954). The particular distribution of the M–N blood groups that people had was explained by reference to the H–W law, since the observed (phenotypic) frequencies were:

M	MN	N
28·38	49·57	22·05

and, if the H–W law were being followed exactly, the (phenotypic) frequencies would have been:

M	MN	N
28·265	49·800	21·935

Differences between the observed and expected frequencies were taken to be irrelevant, given the relatively small sample of people involved (indeed, Race and Sanger, in reporting this study, note that so close an agreement as they have between observation and theory would occur only 1 time in 10).

Obviously, in this case, the authors of the study were relying on some bridge principles. In particular, they were assuming that if a person had M-type blood, then he would have the M gene, and so on; and they were using these to infer from Hardy–Weinberg ratios at the genotypic level to expected and actual ratios at the phenotypic level. Moreover, these authors do actually make explicit the principles bridging genotype and phenotype, and (since the pertinent phenotypic characteristics are not directly observable) they fill in the gap between phenotype and observation. About the genotype–phenotype links they write:

According to the theory there are two allelomorphic [allelic] genes M and N, either of which determines the presence of the equivalent antigen on the red cells.

Genotype	Phenotype or Group
M M	*M*
M N	*M N*
N N	*N*

(Race and Sanger, 1954, 54)

Earlier, they describe agglutination tests for the detection of antigens. 'A serum containing a known antibody is added to a saline suspension of red cells. If the cells carry the equivalent antigen they are agglutinated; if no agglutination occurs it is concluded that the cells lack the antigen' (Race and Sanger, 1954, 3). Hence, we have bridge principles.

Clearly, an exposition of one small part of Mendelian population genetics cannot prove definitively that the whole discipline is axiomatized, and I fully admit that many parts of the discipline proceed in a rather loose and informal way. But of course, the same can be said of most parts of physics. The important point is that it does now seem to be the case that, given that there is nothing forced about my example, the onus is upon biology's detractors to prove that, despite appearances to the contrary, Mendelian population genetics is not a genuine axiomatic system.

3.2   *Natural selection*

The final task left in my programme at this stage of the book involves a consideration of the nature of explanation in Mendelian genetics. In order to give this consideration, it will be worthwhile if first we digress and discuss what is probably the most (philosophically) controversial topic in the whole of biology—natural selection. Having done so, the philosophical discussion can be enriched with a fairly sophisticated example of a genetical explanation (and, incidentally, we shall extend our coverage of the basic concepts and claims of Mendelian population genetics).

As an introduction to the concept of selection, let us start by reflecting on the fact that the H–W law often strikes people as being little more than a truism, namely that if nothing happens to upset a population, then everything (i.e. the gene ratios) will remain the same. In fact, the law is no more of a truism than is Newton's first law of motion, namely that if nothing happens to upset a body's state of rest or motion, then everything will remain the same. Indeed, the H–W law seems important to population genetics for much the same reason that Newton's first law seems important to his mechanics. By adopting his law, the Newtonian has a firm base from which to work. He can introduce factors for change, knowing that they will not be swallowed up in an already-existing, unstable state. In a like manner, the population geneticist can introduce factors causing genetic change, confident that a background stability is guaranteed by the H–W law.

There seem to be two major causes of genetic change considered by the population geneticist: *mutation* and *selection*. Mutation we have already encountered. As we have seen, this change is believed to be random, in the sense that the cause of the change is not a function of the needs of the gene-carriers. But, fortunately, it does seem to be the case that mutation is a fairly regular process—in particular, it would seem that one can quantify the rate of mutation from one gene to another in a fairly unproblematic sort of way.[1] On the other hand, for a non-biologist, the other major cause of gene-ratio-change, natural selection, seems to cause grave difficulties, although the basic idea incorporated in the concept of natural selection—that some organisms are, by virtue of the special characteristics they have, better at the task of surviving and reproducing than other organisms and that thus there is a 'differential reproduction' between organisms, having the consequence that some genes will get passed on in higher proportions than other genes—is not that complex.

However, what the critics of natural selection argue is that all that natural selection states is that some organisms (by definition 'the fitter') survive (and presumably reproduce); but they also argue that since fitness must be defined in terms of survival and reproduction, natural selection reduces to the (empirically) empty tautology that 'those which survive (and reproduce) are those which survive (and reproduce)'. Moreover, the number of critics of this kind is legion. Let me mention here but three. Manser argues that the whole of evolutionary theory is suspect, mainly because it rests on circular concepts. Of natural selection he writes that 'there can be no independent criterion of fitness or adaptability; survival and adaptability or fitness are necessarily connected' (Manser, 1965, 26). Hence, assertions about natural selection cannot rise above the analytic. Barker defends Manser's interpretation, writing that 'what constitutes fitness will obviously vary with the conditions, since there could be no single empirically isolable characteristic or set of characteristics such that any organism which possessed them would in all circumstances survive' (Barker, 1969, 274). Because of this, he argues that natural selection presents us with a real tautology, although, for what it is worth, he does think that it is a 'significant' tautology. And even Smart writes that: 'We can say that even in the

---

[1] Mutation is random in perhaps another way in that we cannot predict when a particular gene will mutate—our predictions must be over groups.

great nebula in Andromeda the "fittest" will survive, but this is to say nothing, for "fittest" has to be defined in terms of survival' (Smart, 1963, 59).

Obviously this is a pretty serious charge, or at least the critics think it is. Every scientific theory contains some *a priori* truths, mathematical truths for instance, and some of science's best known laws hover on the edge of analyticity, Newton's second law of motion for instance. But if a supposed major cause of gene-ratio-change is no more than a hollow tautology, then if we have a theory, for example Mendelian population genetics, which on the one hand uses the notion of selection extensively, but on the other hand claims to tell us significant truths about the world, then Manser's conclusion that such a theory presents 'a *picture* of the process' of hereditable change is charitable. Hence, a first priority for us must be to address ourselves to the question of whether or not natural selection is in fact tautological. But as soon as we do address ourselves to this question, immediately one thing becomes obvious. Whether or not some formulations of the principle of natural selection are tauto-logical, biologists in talking of natural selection are certainly pointing to a non-trivial possible way in which genetic change might be effected. To see this, suppose one had two equal-numbered groups of organisms, that their members interbred within the groups but that the groups were reproductively isolated (from each other as well as from other groups), and that the first group were homozygotes for gene *A*, and the second group were homozygotes for gene *B*. Suppose now that gene carriers *A* had twice the sex drive of gene carriers *B* (by virtue of having gene *A* rather than gene *B*), and suppose as a consequence that whereas the *B* carriers just had one offspring for each parent, the *A* carriers had two. In one generation the *A* carriers would outnumber the *B* carriers 2 to 1. That is to say, whereas there had once been a 1 to 1 ratio between the *A* and *B* genes, the ratio would now be 2 to 1. And this is a conclusion which obviously holds, even though there might be some other causes randomly destroying members of the new generations, so that in absolute numbers, the combined size of the two groups remains the same.

Clearly, to claim (as biologists do) that a phenomenon like this occurs is not to talk in tautologies. Of course, one's claim might be false—one might never get a phenomenon like this in the world, or such a phenomenon, even if existent, might have no biological significance. But this, of course, is a risk that every scientific theory runs. In other words, it would seem that in supposing that there is natural selection, contrary to the claims of the philosophers we have just mentioned, biologists no less than physical scientists are making factual claims about the world.

The question now becomes, why do so many people have so much difficulty over the concept of natural selection? At least part of the trouble is, I think, that the concept is much more subtle than most people recognize. On top of identifying instances of differential reproduction (such as the one I have just discussed), biologists want to have some way of identifying the various kinds of organisms involved in the selection process. Now, as the critics point out, what biologists want to do is, *in some sense*, to label the successful organisms the 'fit' organisms. Furthermore, as the critics also point out, ultimately the only way in which this can be done is by referring to the comparative reproductive successes of different organisms. But it is easy to see that one does *not* have, as the critics frequently claim, a straight identification of those which survive and reproduce with those which are fit. No biologist would want to deny that often it is the unfit organism which reproduces, whereas the fit fails to do so. Darwin himself recognized this when, of his own concept of natural selection, having noted how many eggs and seeds in each generation get destroyed, he wrote that 'many of these eggs or seeds would perhaps, if not destroyed, have yielded individuals better adapted to their conditions of life than any of those which happened to survive' (Darwin, 1959 ed., 173). Similarly, modern evolutionists concede that, through no doing of their own, the weedy little runt often has more offspring than the prize specimen. Indeed, one of the most interesting claims of modern biology is that sometimes, through chance, those with a lower fitness will be more (reproductively) successful than those with a higher fitness, to the extent that the carriers of fitter genes will be eliminated from the group. This claim about 'genetic drift' is highly controversial; but not even its most severe critics claim that it is a contradictory notion, which it would have to be were fitness always directly equated with repro-ductive success (see Wright, 1931; Dobzhansky, 1970; Ford, 1964).

What then is the total relationship between fitness and reproductive success? The situation seems to be something like this. Biologists recognize that, under certain specified circumstances, given a large enough population, by virtue of the particular characteristics which they have (conferred in part by their genes), a higher proportion of one kind of organism do in fact reproduce than another. Those which do have the reproductive edge they call the 'fitter', and they say that the fitter are more successful by virtue of their peculiar variations (which are called 'adaptations'). Now, it should be noted that even here it is not being claimed that *all* of the fitter are the better reproducers. Rather, it is being claimed that given enough organisms a sufficiently high proportion of the fitter are better reproducers, so that the fitter members as a whole have the reproductive edge. Moreover, this claim is indeed analytic, for this is a definition of what is meant by 'fitter', and this kind of stipulative definition is true by fiat. But it should be realized that it is not on this definition that the biologist rests his empirical claims. His basic claim, other than his claim that a differential reproduction does occur, is one which could well be false, for his claim is that there is a certain constancy about which organisms in a given type of situation will turn out to survive and reproduce at a rate better than others. It could logically be the case that in a certain situation that genotype carriers A_1A_1 are much more successful at survival and reproduction and hence fitter than genotype carriers A_2A_2, and that in identical types of situation genotype carriers A_2A_2 are much more successful at survival and reproduction and hence fitter than genotype carriers A_1A_1. If this kind of randomness held, then the biologist's treatment of natural selection would fall to the ground, for he would not then be able to connect up fitness on one occasion with fitness on other occasions. The biologist's argument is that not every logical possibility is realized—he thinks that reproductive success is (in part) a function of the phenotypic characters which are caused (in part) by the genes, and his claim is that, given *any* large enough population of kind P and given *any* environment of type E, a certain assignable proportion (say p) of organisms with phenotype of type A brought about by genotype of type x will survive and reproduce at a specified rate. The biologist thinks that he, like other scientists, can generalize in this sort of way. However, the fact that he then defines those which are the most successful organisms in such environments as the 'fitter' or 'fittest' makes no difference to the empirical nature of his claims, for he might be wrong in his asumption that these fit organisms (from environment to environment) are all of the same kind. Hence, although people like Manser are right in thinking that there is something non-empirical involved here (i.e. a definition), no grounds have been given for assuming that a theory involving selection can give but a picture.

None of what I have written is to deny that in practice it is often very difficult to decide which characteristics are contributing to fitness or to find whether or not different environments are similar in relevant respects. Problems of this nature will be considered in a later chapter. What here concerns us is a theoretical criticism and my reply. For reasons just given, I suggest that, in relying on the concept of selection, the biologist in no way makes his ultimate product inferior to those of the physical sciences; and with this conclusion, let us now turn to the question of the nature of explanation in population genetics. (I defend Darwin's concept of selection against criticisms in Ruse, 1971a.)

3.3 *Explanation and population genetics*

The most famous (or, perhaps, notorious) claim by logical empiricists is that explanation in the physical sciences conforms, or ought to conform, to what has been called the 'covering-law model' of scientific explanation. Essentially, this model has two requirements, for it is argued that in order to 'explain' something, that is, to make something in some way clear or to throw light on it, one must show how the thing had to occur given the way that the world is, in particular, given the way that the world runs along certain regular paths. Hence, first it is argued that an adequate explanation must make reference to at least one law of nature in the thing doing the explaining (the '*explanans*'), and secondly, it is argued that there must be a strong link between the *explanans* and the thing being explained (the '*explanandum*'). Just exactly what constitutes a 'strong link' in this context has been a matter of some debate. Many writers on the subject believe that (a statement about) the *explanandum* must be a *deductive* consequence of (statements about) the *explanans* (e.g. Brodbeck, 1962); however, other writers believe that an *inductive* link between *explanans* and *explanandum* will suffice, so long as the former makes the latter 'highly probable' (e.g. Hempel, 1965). Important though this dispute is, limitation of space prevents me from going into the problem in any depth. Hence, I shall usually rather bluntly refer only to the need of the *explanandum* to be 'deduced' from the *explanans*. To be honest, since every covering-law model demands at least that the *explanans* make the *explanandum* highly probable, I do not think that in this case my riding roughshod over a delicate philosophical problem is too heinous a crime, nor will my action make any significant differences to the conclusions I want to draw about biology. But I do recognize that there is a problem I am avoiding, and neither my silence nor my terminology should be taken to indicate my personal position about the proper solution.

Symbolically, one can lay out the covering-law *schema* as follows:

$$C_1, C_2, \ldots C_m \qquad \textit{Explanans sentences}$$
$$L_1, L_2, \ldots L_n$$

$$\overline{}$$

$$E \qquad \textit{Explanandum sentence}$$

In the *explanans*,[2] the L's stand for laws and there must be at least one of these. The C's stand for statements about initial conditions; but, it should be noted that although C's will occur if the explanation is of a *particular* event or phenomenon, one could have an explana-

[2] From now on I shall usually ignore the difference between the *explanans* or *explanandum* and the statements about these.

tion without any *C*'s at all in the *explanans*. Such an explanation would obtain when the *explanandum* is itself a law. What one would be doing in a case like this would be showing how one law follows from another law. In such an instance, provision of an explanation would seem to be equivalent to showing how part of one's theory could be put in axiomatic form. For this reason, it would seem to me very strange were one to claim that a certain scientific theory is or could be made axiomatic, and yet one were to claim that none of the explanations occurring about the subject matter of that theory are (or could be) covering-law explanations. In confirmation of this point it can be seen without argument that the derivation given earlier of the H–W law, although given to support my claim that population genetics is axiomatic, seems also to support the claim that population genetics gives explanations (in this case of the H–W law) which fit the covering-law model. (Supporters of the covering-law model seem all to agree that the explanations of laws involve deductive inferences, although as we shall see in Chapter 10, this does sometimes presuppose a somewhat idealized view of science.)

However, provision here of another explanation from population genetics which fits the model will not be redundant, for through such a provision can be shown the weakness of another claim by a philosopher who would drive a wedge between biology and the physical sciences. Obviously a critic of biology like Smart, namely one who denies that biology has any laws at all, would deny that population genetics could contain any covering-law explanations. But, at least one other philosophical commentator on biology, one who is much more sympathetic to biology than Smart and one who is even prepared to admit that biology has laws, denies that the covering-law model seems an appropriate model for explanations in the part of biology we are considering. In *The Ascent of Life*, Goudge argues that laws in such explanations serve as 'components in evidential systems which render *explicanda* [i.e. *explananda*] intelligible or rationally credible . . . They constitute evidential grounds for what is to be explained. The situation can be compared with that of a partially finished crossword puzzle in which various entries interlock and offer each other support' (Goudge, 1961, 123–4). Goudge even goes as far as to say that 'the result may be virtually conclusive and leave no room for alternative solutions' (Goudge, 1961, 124); but he still feels that the pattern of explanation to be found in the physical sciences is not that which is to be found in this part of biology, and he claims that 'a crossword puzzle is a more apt model' (Goudge, 1961, 124).

Against Goudge I offer the following phenomenon which I believe is given an explanation of a kind identical to that to be found in the physical sciences and which is a covering-law explanation:

In certain parts of Africa up to 5 per cent of the population dies in infancy or early childhood through a heritable form of anaemia ('sickle-cell' anaemia). Since it is known that the acute anaemia occurs in homozygotes for one particular gene, this points to the fact that up to 40 per cent of the members of the afflicted populations are heterozygotes for the gene, apparently carrying it recessively. Given that there is so severe a pressure against the gene, and given also that it is far too common to be a result solely of mutation to it, it is very strange both that the gene should be so widespread and that it should be so persistent.

The explanation of the magnitude of sickle-cell anaemia is now known. Populations which carry the sickle-cell gene in high proportions are just those populations living in those parts of Africa where a particular malarial parasite is endemic. Moreover, the sickle-cell gene is not entirely recessive—indeed, what it does is to show its effects in the heterozygote by providing an immunity to malaria, an immunity not possessed by the homozygote for the normal gene. Thus selection in favour of the sickle-cell heterozygote 'balances' out the severe selection against the sickle-cell homozygote, and so the gene persists in the population. But how, it might be asked, can something like this occur? The selection against the sickle-cell homozygote is far, far more severe than selection in favour of the heterozygote (over the normal homozygote). After all, by no means all of the normal-gene homozygotes die of malaria. It is at this point that population genetics steps in to complete the explanation of this instance of balanced 'polymorphism' (i.e. the occurrence and persistence of several morphological (phenotypic) forms in the same population). The necessary mathematical steps are both brief and simple, and so let us follow them in detail.

Suppose that we have a population with two kinds of gene (at one locus) as we have in the above instance. Suppose that initially in the population the fraction of one kind of gene (say, A_1's) is q, and that therefore the fraction of the other kind (A_2's) is $1 - q$. Suppose also that the population has been breeding randomly, and that therefore the genotypes are distributed according to the H–W law (i.e. there are q^2 of the A_1A_1 homozygotes, $2q(1 - q)$ of the A_1A_2's, and $(1 - q)^2$ of the A_2A_2's). Now, let us assume that the population shares another of the features found in the above example, namely that the heterozygote is fitter than either homozygote. We can quantify this assumption by assuming that for every A_1A_2 that survives (to reproduce), proportionately, S_1 of the A_1A_1's will die (without reproducing) and S_2 of the A_2A_2's will similarly die. Now, thanks to the H–W law, we can calculate the proportions of genotypes we should expect to find after one generation of such selective pressures. For clarity, we can draw up a table as follows:

Genotype	A_1A_1	A_1A_2	A_2A_2	Total population
Adaptive value	$1 - S_1$	1	$1 - S_2$	
Initial frequency	q^2	$2q(1-q)$	$(1-q)^2$	1
Frequency after the selection	$q^2(1 - S_1)$	$2q(1-q)$	$(1-q)^2(1-S_2)$	$1 - S_1q^2 - S_2(1-q)^2$

This means that the rate of change, Δq, of the frequency of the A_1 genes in the population in one generation is:

$$\Delta q = \frac{q(1-q)\,[S_2(1-q)\,-\,S_1q]}{1\,-\,S_1q^2\,-\,S_2(1-q)^2}$$

Next, let us recall that in the kind of situation which we have above, the polymorphism is balanced. Therefore, $\Delta q = 0$. Hence, solving for q, we get

$$q = S_2/(S_1 + S_2).$$

Finally, let us supply some actual intensities for the different selections. Notice how there is nothing tautological about making these assumptions about the intensities and assuming that they hold from generation to generation, which there would be were the critics of selection correct. Indeed, so involved an empirical matter are these assumptions, that I shall take the liberty of using round figures within the limits of error known for real-life cases, so that the inferences can be seen more clearly. (These are figures used by Livingstone, 1971.) Let us, for simplicity, suppose that in the case of sickle-cell anaemia all of the sickle-cell homozygotes die without reproducing. Hence, relative to the heterozygotes none reproduce, and thus, for them, $S_2 = 1$ (i.e. $1 - S_2 = 0$). Suppose also that, in some population, for every four heterozygotes that reproduce only three normal-gene homozygotes reproduce, the other dying of malaria. This means that $S_1 = \frac{1}{4}$. Hence, if A_1 is the normal gene, and A_2 is the sickle-cell gene,

$$q = \frac{1}{\frac{1}{4} + 1} = \tfrac{4}{5}.$$

This means that in *every generation*, $\tfrac{4}{5}$ of the genes will be A_1 and $\tfrac{1}{5}$ will be A_2, even though the A_2 gene kills its homozygotes before reproduction. Moreover, *in every generation* the gene carriers will be distributed as follows:

A_1A_1	A_1A_2	A_2A_2
$\frac{4}{5}^2$	$2 \cdot \frac{4}{5} \cdot (1 - \frac{4}{5})$	$(1 - \frac{4}{5})^2$

This means that for every hundred members of the population, approximately 64 will be A_1A_1 homozygotes, 32 will be A_1A_2 heterozygotes, and 4 will be A_2A_2 homozygotes and will die without reproducing. This is a condition which will continue indefinitely until something changes.

We have now a full explanation of the persistence of a phenomenon like sickle-cell anaemia, and although I have used round figures in my discussion, these fall within the limits known for real-life instances (see Raper, 1960; Livingstone, 1967; 1971). Moreover, I suggest that we have here before us, as I claimed, an example of a covering-law explanation of a particular phenomenon. One's *explanandum* is the way in which the sickle-cell gene persists from generation to generation, causing a certain proportion of each generation to die young from anaemia (a proportion which, in my example, I put at 4 per cent). The *explanans* contained laws, particularly the H–W law, statements about particular conditions, for example that the gene-ratio-change was zero and about the particular values of selective pressures against the homozygotes, and the statement about the *explanandum* followed from these. In short, I would suggest that, granting that this explanation is typical of population genetics (and I see no reason to doubt this), it seems proper to conclude that the explanations of population genetics share the same form as do those of the physical sciences. Contrary to Goudge's claim, there is no need to invent a new model—the 'crossword' model of scientific explanation—to account for explanations in this part of biology.

This now concludes my exclusive concern with genetics. Admitting that much has had to be taken for granted, I think that it has been shown that many of the claims made about the physical sciences have a corresponding truth in Mendelian genetics. Hence, I shall turn now to a far more controversial area of biology—evolutionary theory.

4

THE THEORY OF EVOLUTION

I: THE STRUCTURE

The reader who has some knowledge of modern evolutionary theory will possibly, by now, be feeling somewhat disoriented, not to say irritated. He will know, to put it very briefly, that the modern theory of evolution is simply the working out of the claim that organisms did not just 'spring into being' but were rather the result of a long, relatively slow process of natural selection acting on constantly appearing, random, heritable variations. But this, such a reader might with justice point out, seems suspiciously similar to the theory I have just been considering under the heading of 'population genetics'. Where then is the place of evolutionary theory? Is 'population genetics' just another name for it, in which case we seem already to have covered the theory, or are the two things quite separate, in which case (assuming that evolutionary theory is in fact about selection and so on), how do we account for the fact that population genetics and evolutionary theory seemingly share so many concepts?

A reader's bemusement at this stage is quite understandable; but, I would suggest that neither alternative points in quite the true direction. To find the right answer about the nature of modern evolutionary theory (and its relation to population genetics) one must understand that basic to modern evolutionary thought is the claim that the answers to questions about large-scale evolutionary changes are to be found in our knowledge of small-scale evolutionary changes—changes so small that they would not normally be labelled 'evolutionary'. Modern biologists believe that the organic world which we see around us (and, of which, we are a part) is indeed the product of a slow, gradual, evolutionary process; however, they believe that the process which brings about the largest changes is no more than the long-term cumulative effect of processes which bring about the smallest heritable changes. But, since population genetics is the science which studies these small changes, we can therefore see its importance for the study of large changes—the study which is called 'evolutionary theory'. *Population genetics is presupposed by all other evolutionary studies.*

Let us consider this point I am making in more detail. As we shall see in this and the succeeding chapters, many (if not most) of the misconceptions of evolutionary theory stem from the failure to recognize the fundamental importance of population genetics for evolutionary thought. (*Loci classici* on evolutionary theory include Dobzhansky 1937, 3rd ed. 1951, revised and retitled 1970; Simpson, 1953; Mayr 1942, 1963; Stebbins 1950.)

4.1 *The structure of the theory of evolution*

A great many different areas fall under what are loosely called 'evolutionary studies'. There is *systematics*, the study of the distribution of organisms and the reasons why they fall into the particular kinds of groups that they do. There is *morphology*, the study of the different kinds of characteristics that organisms have, together with the theorizing about the reasons for such characteristics. There is *embryology*, the study of the development of organisms. There is, of course, *paleontology*, the study of organisms long dead and fossilized. And there are many other areas. All of these disciplines have factual claims peculiar to themselves—for example, embryology is based in part upon claims about the needs of embryos and the particular factors influencing development. Also, many of the disciplines borrow from other disciplines, particularly using as unproven assumptions the conclusions derived in other areas. For example, the paleontologist will rely on the work of the morphologist to support his inferences that certain fossil structures represent characteristics which had certain functions for their possessors. However, what I suggest is that all the different disciplines are unified in that they presuppose a background knowledge of genetics, particularly population genetics. The knowledge that the population geneticist supplies about the way in which heritable variations are transmitted from one generation of a population to the next is presupposed and drawn upon by every kind of evolutionist, even (nay, particularly) those like the paleontologist who study the largest of heritable changes, from fish to reptiles, and from reptiles to mammals and birds.

Figure 4.1 will perhaps make a little clearer the structure I am supposing evolutionary theory to have.

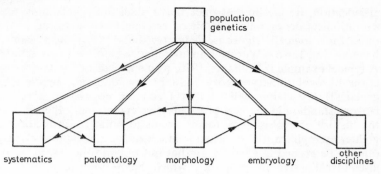

Figure 4.1

(In this figure, the rectangles represent various disciplines; the double lines the links between population genetics and other areas—such links are actually supposed to exist; and the single lines links between the subsidiary disciplines—although such links do exist, those shown in the figure are just illustrative, they do not necessarily denote particular instances.)

Given the approach adopted so far in this book, the obvious question which arises is in just what respects evolutionary theory (in the way conceived above) is a theory like physics and chemistry? The answer to this is that, by virtue of the fact that the theory of evolution has population genetics as its core, it shares many of the features of the physical sciences. The most vital part of the theory is axiomatized; through this part (if through nothing else) the theory contains reference to theoretical (non-observable, etc.) entities as well as to non-theoretical (observable, etc.) entities; there are bridge principles; and so on. However, once this has been said, it cannot be denied that the *whole* theory does not possess the deductive completeness possessed, say, by Newtonian mechanics. Because of many factors—the newness of the theory, the fact that many pertinent pieces of information are irretrievably lost, the incredible magnitude and complexity of the problems, and so on—many of the parts of evolutionary theory are just 'sketched in'. What one finds, instead of explicitly formulated strong links between various levels, are suggestions, hypotheses, extremely weak inductive inferences, and frequently, outright guesses. Hence, at best, one can say that evolutionists have the hypothetico-deductive model as an ideal in some sense—they are far from having it as a realized actuality.

Of course, the next question to come up is in just precisely what sense can one say that evolutionists have the hypothetico-deductive model as an 'ideal'? I hope that in this and the next chapter my usage of this term will become fairly clear; but towards a preliminary clarification, let me note at once that I shall not be claiming that evolutionists do or must put all other considerations aside in the quest for a logically rigorous, deductive theory. (For this reason, I have no time for the Woodgerian 'axiomatize before-all-else' school. A typical example is Williams, 1970, who succeeds in her axiomatization of evolutionary theory only by avoiding all mention of genetics!) My claims will be that I can see no good theoretical reasons standing in the way of a deductive evolutionary theory, that the history of evolutionary theory seems to point towards an ever greater exemplification of the axiomatic ideal, and, most importantly (and to be argued in Chapter 5), that not only do there seem to be grave weaknesses in proposals which have been forwarded as alternatives to the idea of evolutionary arguments as deductive (or perhaps strong inductive) sketches, but that the only way of remedying these weaknesses seems to be one pointing in the direction of completed deductive arguments as the ultimate realization of evolutionary studies.

However, before we explore these points in more detail it will be convenient first to examine other aspects of the way in which I have characterized evolutionary theory, for, as we have seen, my claim is not merely that the hypothetico-deductive model is relevant as an ideal for evolutionary theory, but my claim is also that evolutionary theory is a *unified* theory with population genetics as its presupposed *core*. Since one of the most intelligent and well-informed of all philosophical writers about biology, Beckner, has presented a conception of evolutionary theory which differs from my conception, specifically with respect to these latter claims that I have made, much (philosophical) light will be thrown on evolutionary theory by an examination of our rival views. Therefore, to conclude this section I shall look briefly at Beckner's rival analysis of the theory. Then, in the next two sections, I shall compare my account with Beckner's, and in so doing I shall hope not merely to refute Beckner but shall try to illustrate in much more detail some of my claims about the theory. Finally, to finish the chapter, I shall consider three arguments by Goudge purporting to show that the axiomatic deductive ideal is inappropriate for evolutionary theory. As I have said, some of the really difficult questions as to whether or not the deductive ideal is in any sense appropriate for evolutionary studies will be left until the next chapter.

Beckner writes about evolutionary theory as follows:

... if we look in evolutionary theory for the pattern of theoretical explanation exemplified in that paradigm of theory formation, Newton's explanation of Galileo's and Kepler's laws, we shall be disappointed. Evolution theory does not attain its ends by exhibiting, e.g. Williston's and

Bergmann's principles as consequences of one or more hypotheses of greater generality. There are a number of small hierarchies of this character scattered about evolution theory, but the theory as a whole does not approach this type of organization. It is impossible to be dogmatic on the point, but it does seem to be true that this fact is not due to the under-developed state of biology, but to the nature of biological subject matter. ... My own view is that evolution theory consists of a family of related models; that most evolutionary explanations are based upon assumptions that, in the individual case, are not highly confirmed; but that the various models in the theory provide evidential support for their neighbors. The subsidiary hypotheses and assumptions that are made for the sake of particular explanations in one model recur again and again in other related models, with or without modification and local adaptation. To use the metaphor of Agnes Arber, biological theory is less 'linear' than e.g. physical theory, and is more 'reticulate'. (Beckner, 1959, 159–60)

If one were to draw a picture representing the way in which Beckner sees evolutionary theory, then it would come out something like Figure 4.2.

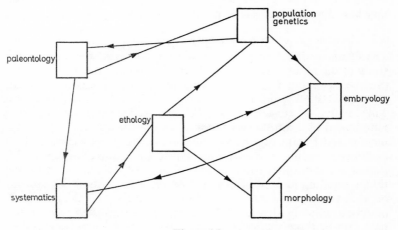

Figure 4.2
(No attempt has been made at completeness, and all of the arrowed lines are merely illustrative.)

Now clearly it will not be a particularly easy task to decide between the way in which I view evolutionary theory and the way in which Beckner sees it. Apart from anything else, there is quite a bit of overlap in the two visions, for we both recognize that although there are formal parts, much of the theory is diffuse and merely suggestive. Nevertheless, one can, I think, give arguments showing that my characterization of evolutionary theory, one which gives a special

role to population genetics, is better than Beckner's. I shall here offer
two such arguments. The first argument will be in the form of an
analysis of a type of explanation made, in an area of evolutionary
studies, with the aid of evolutionary theory. I shall show the role and
importance of population genetics in this explanation, citing it as
evidence in favour of my viewpoint. Admittedly, Beckner does
concede that things like the H–W law keep cropping up in evolutionary explanations and he even admits the great value of population
genetics; but for him, this seems to be a matter of mere chance or, at
least, as something which does not conflict with his above-given
characterization of the theory—a characterization which deliberately
does *not* put any area of evolutionary studies in a unique position
above all others. Through my example, I shall try to show that it is
better to view the occurrence of such laws as the H–W law as
pointing to fundamental aspects of evolutionary theory. My second
argument will involve a discussion of the one positive reason that
Beckner offers in favour of his view—the supposed non-connection
of evolutionary laws with the main part of the theory. I shall argue
that here also my analysis of the theory is more satisfactory.

4.2   *An example of an evolutionary explanation*

The example that I want to consider comes from *systematics*, the
study of the nature and the distribution of animal and plant groups.
(Strictly speaking I shall be giving a number of examples since,
although I shall speak of 'the explanation', in fact my illustration
will involve a cluster of very closely related things all being
illuminated by evolutionary theory.) The particular kind of phenomenon that I want to discuss and around which my example of an
evolutionary explanation revolves is the distribution of organisms on
islands some distance from the shores of large land-masses, and the
things standing in need of explanation seem to be these. Organisms
on such islands are often similar to the organisms on the mainland,
but not absolutely identical. Moreover, the islands themselves often
have two or three species of a particular genus of organism, and so
again we have organisms which are similar to each other (and to the
mainland organisms) but not identical. Furthermore, the organisms
which have this kind of similarity are usually things like birds,
reptiles, and insects. Oceanic islands rarely have organisms like
mammals, and when they do, they are often very different from those
on the mainland. The most famous example of this kind of phenomenon is Darwin's finches, a subfamily (*Geospizinae*) of the finch
family (*Fringillidae*). These finches are grouped into four genera and
fourteen species, and they occur on the Galapagos Archipelago, a
group of islands in the Pacific, some 600 miles west of Ecuador. As

is well known, recognition that so small a total land mass carries so many different forms of what is essentially one kind of bird was one of the chief factors triggering Darwin's quest for, and discovery of, a theory of evolution.

The explanation that evolutionists give of this and like phenomena begins as follows. First, they argue that in such cases the organic similarity presupposes a group of ancestral mainland organisms, from which both the (modern) mainland and the island populations are descended. They also argue that the passage between mainland and island (and island and island) is made infrequently (because of natural barriers like winds), but that if it occurs at intervals, the new arrivals will evolve into a new species, similar to but different from those already there. Thirdly, they argue that the kinds of organisms which will settle on islands obviously must be those with some chance of getting there. Birds, for example, have such a chance of being blown there by the wind. Large land animals, on the other hand, have little or no chance of crossing comparatively large tracts of water.

As an illustration of an explanation such as this it is worth considering the explanation of Darwin's finches given by David Lack in his monograph, *Darwin's Finches*. Lack points to the barriers formed by the water between the islands of the archipelago, noting that birds are often reluctant to cross watery expanses, even though they are physiologically capable of doing so. He also tries to show why the finches were able to evolve as much as they did, putting this down mainly to the absence of other passerine-bird competitors on the islands. Lack even points to the fact that the degree of speciation seems to be a direct function of the relative isolation within the island group of the island on which a bird population occurs—the moderately isolated islands having a lower proportion of peculiar forms than the most isolated islands, but a higher proportion than the central Galapagos islands. Figure 4.3, giving the percentage of endemic forms of Darwin's finches on each island, shows clearly the effect of this isolation. Lack adds that, from the bird viewpoint, Charles, Chatham and Hood, are more isolated than the map might suggest, since the trade-winds blow from the south and south-east, making bird dispersal comparatively difficult.

However, illuminating as all of this may be, as one reads evolutionists' discussions of phenomena like Darwin's finches, it rapidly becomes clear that they think that reasons of the sort just given provide only part of the explanation of the peculiarities of island organisms. Evolutionists want to show just why it should be that a group of organisms, isolated from the mainland parental group, evolve into new species. Why, for example, even under the

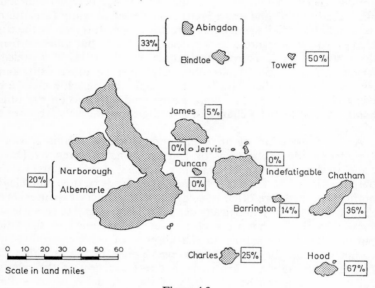

Figure 4.3
Percentage of endemic forms of Darwin's finches on each island
showing effect of isolation. (From Lack, *Darwin's Finches*, Cambridge
University Press.)

new island conditions encountered by the founding organisms should
evolution occur at all? It is at this point that genetical considerations
become paramount and, it should be noted, evolutionists state this
fact quite explicitly. For example, Dobzhansky at the beginning of
his very important work, *Genetics and the Origin of Species*, writes
that:

Evolution is a change in the genetic composition of populations. The study
of mechanisms of evolution falls within the province of population
genetics. Of course, changes observed in populations may be of different
orders of magnitude. Experience shows, however, that there is no way
toward understanding of the mechanisms of macroevolutionary changes,

which require time on geological scales, other than through understanding of microevolutionary processes observable within the span of a human lifetime, often controlled by man's will, and sometimes reproducible in laboratory experiments. (Dobzhansky, 1951, 16)

In the case of something like island species, the way in which this appeal to population genetics is translated into practice seems to be basically through the realization that under the very different conditions that prevail on islands (as opposed to the mainland) the new founding populations will be subject to severe selective forces of a kind quite different from those encountered by the mainland parential organisms. In other words, a differential reproduction will occur within the island groups; but since conditions differ from the mainland (in the case of Darwin's finches, the absence of passerine-bird competitors on the archipelago is one major difference in environmental conditions), the fit island organisms will tend to be fit by virtue of phenotypic characteristics which are not necessarily those conferring fitness on mainland organisms. These phenotypic characteristics are, as we have seen in the last two chapters, thought to be correlated in some way with genotypes, and thus, through selection (and new mutation) we get a divergence of the island populations from the mainland population.

But notice how this supposition of a divergence, even if we do (through bridge principles) assume a correlation between genotype and phenotype, still requires assumptions like Mendel's laws—if such laws did not hold, and were not taken as the starting-point by evolutionists, then there would be no reason for concluding that even the most severe selective forces would leave their marks on future generations. If, for example, genes in some way blended in each generation (and this was reflected in a blending of phenotypic characteristics), then however advantageous a phenotypic character-istic proved to be, there could be no guarantee that this characteristic could be selected for the future. With blending, in one generation even the most advantageous new phenotypic characteristic would have its effectiveness halved, and in a few generations it would be blended practically out of existence, unless one assumes massive mutation rates counteracting the effects of blending. But such muta-tion rates are known not to hold. Thus we can truly say that it is only because of their knowledge of Mendelian genetics, par-ticularly of Mendelian population genetics, that biologists can be sure that the island selective forces will have evolutionary conse-quences. Hence it is not for nothing that Dobzhansky has written of the Hardy–Weinberg law that it 'is the foundation of population genetics and of modern evolutionary theory' (1953, 53). Without this and like laws, evolutionary explanations would come to naught.[1]

Of course, actually applying with any precision the results of theoretical population genetical studies to large-scale evolutionary changes is, because so many unknown factors are involved, incredibly difficult—hence the need for this philosophical discussion! But some results do bear directly on such changes, as those which have occurred in Darwin's finches. For example, it can be shown that if we had a gene in one of the founding populations of finches which, although recessive, caused a phenotype (in its homozygotes) which had no chance of surviving and reproducing in the new surroundings, then the number of generations (t) for the gene to be reduced in frequency from q_o to q_t is given by the equation (derived, in part, from the H–W law)

$$t = \frac{1}{q_t} - \frac{1}{q_o}.$$

In other words, even if the initial frequency of the gene were as high as 0·5, in a thousand generations it would be practically eliminated from the population. Obviously, selection will often not be as intense as this—but then, compensating for this fact are other facts, namely that we have many more than a thousand generations at our disposal and that if a gene were not completely recessive, the effectiveness of selection would be increased accordingly.

The main outlines of the way in which evolutionary biologists explain phenomena like island populations are now before us. Obviously no deductive completeness has been forthcoming; but we can see how evolutionary phenomena are in some sense explained by assumptions, chief amongst which is the theory of population genetics. In fact, I think the importance of population genetics for evolutionary investigations is even greater than so far indicated, for, as the kind of example we are here discussing amply illustrates, evolutionists use population genetics not merely, as it were, to provide the main force behind evolutionary explanations, but also to flesh out a number of supporting details. Let us take up this matter (through our example) a little more closely.

In the case of island populations it is argued by many evolutionists (e.g. Mayr, 1963) that organisms are given a push in the direction of evolutionary change by what is called the *founder principle*.

---

[1] I am not here arguing that (logically) necessarily evolutionists must use Mendel's laws—given what has happened to the status of Newtonian mechanics in the past century one would be rash to argue this. What I am arguing is that if evolutionists would explain they must use laws of genetics, and for those who accept the synthetic theory then these laws will (with qualifications to be made in later chapters) start with Mendel's laws.

Population genetical studies of groups of interbreeding organisms have shown the, perhaps surprising, conclusion that such groups can be quite variable genetically. Despite a pretty fundamental genetic similarity between the members of such a group, considerable latitude for difference still exists. The phenomenon of balanced polymorphism, discussed in the previous chapter, is one of the reasons that population genetics can give to account for this fact. In the case of balanced polymorphism, as we have seen, given certain (constant) selective forces, carriers of different (competing) genotypes can exist indefinitely within a population. But this genetic variability, highlighted and explained by population genetics, means that founders of island populations will not be 'average' members of their parental species, for there are no average members in quite this way. The founders will carry genes peculiar to a few members of the parental group, and will not carry many of the genes carried by the parental group. This fact is thought to account for the lack of variability commonly found in island groups and, more importantly, for a good part of the evolution of the island groups from the parents and from successive founders (who will in turn form their own inter-breeding groups). A group with small genetic variability facing new selective pressures might well 'improvise' in a way that a large variable group (e.g. the parental group) would not. In the large group, a genotype A_1A_1 might be fitter than a genotype A_2A_2; but in the small group, only the genotype A_2A_2 might exist. But possession of A_2A_2 might mean that genes at other loci are fitter (than they would be if the carrier had A_1A_1), and so on and on, leading rapidly to alteration of the genotypes, drastically different from those possessed by the parental organisms, from those of subsequent founders (who would also be subject to great change), and even from the original founders. Thus we have an additional reason to account for the kinds of organisms found on islands.

Enough has now been said to underline my claim about the importance of population genetics for evolutionary explanations. Given the way that, despite their use of principles particular to the area (e.g. about the ability of birds to cross open water) evolutionists appeal to a background knowledge of population genetics to provide much of their explanation, I find my account of evolutionary theory —one which squarely accepts (and even demands) this use of population genetics—to be more satisfying than that of Beckner. Possibly the reply that Beckner might make at this point is that one example cannot prove my general claims about the structure of evolutionary theory, since my claims are about the way in which population genetics is presupposed by and integrates the *whole* of evolutionary theory. By discussing one example from systematics I have hardly considered all of the theory. There is a little truth in this, although I would suggest that my example is sufficiently typical of evolutionary explanations that one can generalize to other areas of evolutionary study. Indeed, it seems to me that if one adopts a position like Beckner's, one must at least allow the possibility of an area of evolutionary discussion which does not in any way rely on population genetics (even though most may do), and I myself must confess that I cannot see how a discussion which has no direct connexion with the laws of heredity as applied to groups or with natural selection could nevertheless be part of the synthetic theory of evolution. This claim presupposes, of course, that reference to the laws of heredity as applied to groups and to natural selection involve a reference to population genetics. It is a tautology that the laws are part of population genetics. As far as natural selection is concerned, although admittedly Darwin's natural selection operated at the phenotypic level, it is just a matter of fact that the modern conception of selection operates essentially at the genotypic level. Doubters should read the works given on page 48.[2]

Unfortunately, limitations of space prevent my giving several other examples here to illustrate and strengthen my claim about the structure of evolutionary theory; but as I turn now to criticize Beckner's sole argument in support of his position, I would suggest that, from a positive viewpoint, in showing the rightful place of evolutionary laws evidence is provided of the importance and integrating function of population genetics. Moreover, the same point will apply with even more force when we come in the next chapter to consider examples (e.g. from paleontology) which purport

[2] A word of clarification is needed here. Biologists do frequently talk of selection operating at different 'levels'. Thus, for example, Lewontin in his excellent review (1970), distinguishes (amongst others) *gametic* selection, 'the differential motility, viability, and probability of fertilization of gametes that arises from their own haploid genotype', *individual* selection, 'the force on the genetic composition of a population that arises directly from differences in the age-specific mortality and fertility schedules of different genotypes', and *kin* selection, where some individuals of a family (e.g. worker ants) sacrifice their own fitness for the sake of other members of the family. But these distinctions seem not to deny the claim that essentially selection is a genetic process—the different levels indicate the different ways in which the effects of the genes are felt. The one exception to a genetic selection would seem to be some kind of molecular selection, operating at a level below that of the gene (Lewontin, 1970, 2–3). However, at the risk of appearing to defend my thesis by fiat, I would suggest that this kind of selection is outside of the provenance of the *biological* theory of evolution, although this is not to deny that at some point one might develop a *chemical* theory of evolution involving molecular selection and perhaps, in some way, having the biological theory as a special case. These sorts of possibilities will be discussed in Chapter 10; but see also Schaffner (1969c).

to show that the deductive ideal is misplaced in evolutionary thought. Thus Beckner's fears on this score will be allayed.

4.3 Evolutionary 'laws'

Early evolutionists (and some later ones) eagerly enshrined every regularity (real or apparent) which they noticed in evolutionary phenomena in a 'law'. Some of these 'laws' supposedly refer to all kinds of organisms, and amongst them we can mention the two named by Beckner, Williston's principle and Bergmann's principle. The former states that repetitive serial structures in animals evolve so as to become less numerous but more differentiated; the latter states that in colder regions the members of the geographical races of warm blooded animals are larger than the members of the races of the same species in warmer regions. Other similar 'laws' are Allen's rule, that the extremities of warm blooded organisms are smaller in colder regions than in warmer regions, and the 'law' which is probably most famous of them all, Dollo's law, stating that evolution cannot be reversed. Today, possibly the most distinguished advocate of evolutionary laws is Rensch, who makes his claims, as many before him did not, in the full knowledge of contemporary evolutionary theory. Rensch argues eloquently that one can also justifiably locate lawlike trends restricted to particular groups of organisms, and one example amongst many which he gives is that 'animals, the enemies of which find their prey by the eyes, develop protecting colors or shapes (including threatening colors and mimicry)' (Rensch, 1960, 108). Beckner argues that the existence of laws such as these show that evolutionary theory cannot have the same structure as a physical theory (specifically Newtonian mechanics), because they are not exhibited as consequences of evolutionary theory in the same way that Kepler's laws follow from Newton's theory. The question we must ask is whether or not Beckner is right in this claim (and if so, what follows from it).

The first thing one should note is that actually it is to be hoped that most of these 'laws' do *not* follow too precisely from evolutionary theory, because they have so many exceptions. For example, Rensch found 20–30 per cent of palearctic and nearctic birds were exceptions to Bergmann's principle. For palearctic and nearctic mammals the exceptions rose to 30–40 per cent. For this reason, most evolutionists prefer to think of the 'laws' as 'rules' or 'principles', rather than as anything rigidly nomically necessary. Indeed, some evolutionists, Simpson (1963a) in particular, refuse to allow that they are in any sense nomically necessary. Whatever the case may be, no evolutionist gives them a very important role in his studies—certainly nothing like the place accorded to Galileo's or Kepler's laws. They seem to be regarded as rough guides, indicating the need for (and worth of) further analysis.

The second point which should be made about the rules of evolution is that it is not strictly correct to imply that evolutionary theory can throw no light on their truth. The theory certainly does not formally imply them; but the theory (particularly the population genetical part) can give good reasons to show why many of them hold. Indeed, it can even show why one might get exceptions. Thus perhaps the best analogy to these 'laws' is something like Boyle's law, which follows from gas theory if one makes enough (false) simplifying assumptions—these assumptions showing why real gases do not obey Boyle's law. (Although even this analogy is really to give the rules more status than they deserve.) The theory of evolution cannot of course assign a definite percentage to the number of exceptions to the evolutionary rules—for this reason it is perhaps better to regard them as 'loose' laws (or, as Helmer and Rescher (1959) speak of historical generalizations, as 'quasi-laws') rather than as statistical laws. The whole point about a statistical law like the H–W law is that the statistical part is just as necessary as the rest of the law—following from the H–W law is not only the fact that there will be three genotypes in future generations (say, A_1A_1, A_1A_2, A_2A_2), but also the fact that the distribution will be $p^2 A_1A_1 : 2pq A_1A_2 : q^2 A_2A_2$. In the case of evolutionary laws, as we shall see, no definite percentages follow from evolutionary theory. We cannot, for example, infer that exactly 30 per cent of nearctic mammals will be exceptions to Bergmann's principle.

In order to see the relation between evolutionary theory and the evolutionary laws let us take Bergmann's principle, which, it will be remembered, was actually one of Beckner's examples. That something like this should be found to hold is quite intelligible given evolutionary theory, since selective pressures in cold regions will likely differ from selective pressures in hot regions. In particular, for warm-blooded animals in cold regions there will be a greater pressure towards genes causing heat-conservation. But, whilst an increase in size increases the overall body area (and thus gives more scope for heat loss), an increase in size *decreases* the body-volume/surface-area ratio (since the former increases with the cube of the length, whereas the latter merely increases with the square of the length). Hence, it would seem that, overall, the larger animal is more efficient from the viewpoint of heat-conservation, and because of this, the basic principles of selection theory can throw light on why Bergmann's principle holds roughly. It can also illuminate the existence of exceptions. There are several ways of achieving the same end (heat conservation)—one alternative is to make the body-covering thicker.

Sometimes, one could have strong selective pressures in cold regions against genes causing increased body-size, and hence the organism would seek other methods of conserving heat. For instance, one species of pocket gophers has smaller members in colder regions. Apparently, this is due to the fact that its colder regions coincide with higher altitude regions, and that since the soil is shallower in such regions, there is a selective premium for smaller gophers needing less room in which to burrow (Mayr, 1942, 1963).

The other evolutionary laws seem to bear a similar relationship to evolutionary theory. Take, for example, Dollo's law. This is perhaps the most reliable of all of the evolutionary laws; but even it has exceptions of a kind—back mutations occur, and lost combinations of genes can be found. It is even possible, in a way, for organisms to reverse large-scale, evolutionary phenomena—organisms in some lines got smaller and then larger, nearly all the animals without sex have early ancestors without sex and more recent ancestors with sex, and, as is well known, the ancestors of the whale left the sea and then returned. On the other hand, even though exceptions like these are enough to make Simpson refuse to grant that it is a real law, clearly the 'law' does have some plausibility—whales did not go back to being fish, for example. Moreover, Simpson himself justifies the law by an appeal to genetics. He writes:

The statistical probability of a complete reversal or of essential reversal to a very remote condition is extremely small. Functional, adapted organisms noticeably different in structure have different genetic systems. They differ in tens, hundreds, or thousands of genes, and such genes as are the same have different modifiers and are fitted into differently integrated genetic backgrounds. A single structure is likely to be affected by many different genetic factors, and practically certain to be if its development is at all complex. The chances that the whole system will revert to that of a distinctly different ancestor, or even that this will happen for any one structure of moderate complexity, are infinitesimally small. (Simpson, 1953, 311)

Finally, it is worth noting that Simpson concludes this passage by appealing to a technical paper by the geneticist Muller (1939) for support.

I must confess that, dubious though the value of the 'laws of evolution' undoubtedly is, in the light of evidence such as we have just seen, I cannot see that Beckner strengthens his hand by appealing to them for support. They seem to follow in some way from evolutionary theory—they are not isolated statements as Beckner's account implies—and I think that the spelling out of the detail about their relationship to the theory ultimately involves reference to genetics, particularly population genetics. Certainly population genetics would be needed in, say, an explanation of Bergmann's principle which showed how the selection of genes for increased efficiency in heat-conservation could spread such genes through an entire population. Hence, 'laws' like these do seem to be 'consequences of one or more hypotheses of greater generality', for an understanding of something like the Hardy–Weinberg law is required in an explanation of something like Bergmann's principle. Of course, whether a full spelling out of the consequence-relationship would eventually be deductive has not yet been proven; but before I turn to this question, I want first to discuss three rather negative arguments, all purporting to show that such a deductive spelling out would be impossible.

4.4 *Is an axiomatic evolutionary theory theoretically impossible?*

One of the major themes of Goudge's *The Ascent of Life* is that evolutionary theory is a theory of quite a different nature from the theories of physics and chemistry. We have already met one of Goudge's arguments, that which is directed against population-genetical explanations, and in Chapter 5 we shall meet similar arguments. At this point, I want to consider three arguments by Goudge purporting to show that evolutionary theory does not have, and never can have, a (deductive) axiomatic form.

Goudge's first argument runs as follows:

The concepts which occur in a formalized axiomatic system have an absolutely fixed and precise meaning. In this respect they differ from concepts which refer to empirical facts. The latter concepts have an unavoidable imprecision, due to what has been called their 'open texture'. The more complex the facts to which they refer, the greater will the imprecision be. Now the concepts which belong to evolutionary theory refer to facts whose complexity is enormous. Hence to treat these concepts as though they were absolutely precise by incorporating them in a formal system, would be to distort them beyond recognition. (Goudge, 1961, 16)

It is easy to see that, given its present form, this argument is altogether too powerful for Goudge's purpose. If what Goudge writes were true, then not only would biological concepts be excluded from the realm of possible axiomatization, but so also would be the concepts of physics and chemistry. Physico-chemical concepts refer to 'empirical' facts like planets and pendulums and acids; but given Goudge's conclusion, these concepts could never be incorporated within axiomatic systems. Since they obviously can be, his argument must be modified somewhat. Presumably, what Goudge really means is something to the effect that the concepts in an axiomatic system, whether they refer to empirical things or not, must have a fairly limited meaning, but that due to the complexity of evolutionary phenomena, evolutionary concepts could never have such a limited meaning, and therefore evolutionary concepts could never be incorporated within an axiomatic system.

It is not that difficult to see that even this modified criticism is gravely in error. For a moment, let us grant the rather dubious truth of the premise that the concepts of an axiomatic system must have a fairly limited meaning. Goudge's criticism still flounders on the fact that we encounter physical phenomena of enormous complexity —at least as complex as anything biological—and yet concepts which refer to these phenomena can be incorporated within the axiomatic systems of physics. Consider, for example, planets. These are highly complex phenomena—indeed, presumably a full description of what we meant by the planet 'Earth' would involve a full description of all the life on it—but despite this complexity, laws about the planets can be deduced from that axiomatic system known as 'Newtonian mechanics'. Goudge must therefore be mistaken when, *a priori*, he rules out the possibility of axiomatic evolutionary systems because of the complexity of evolutionary phenomena. Physical phenomena are just as complex; but physics has axiomatic systems. Of course, this is not to deny the difficulty of giving an axiomatic theory about complex phenomena; but complexity in itself cannot (for theoretical reasons) bar such a theory.

Goudge's second argument purporting to show the impossibility of ever having an axiomatic evolutionary theory runs as follows:

The framework of statements constituting evolutionary theory is at present lacking in the tidiness and completeness which should exist before any axiomatization is undertaken. Since the theory is expounded discursively by biologists, it is often formulated in different ways with differences of emphasis on details. Modifications are being periodically introduced as fresh evidence comes to light and new theoretical interpretations are proposed. Any attempt to 'freeze' such a growing body of ideas in a formal scheme would result in grave misrepresentations. (Goudge, 1961, 16)

I think that of all Goudge's arguments this is the one with the greatest element of truth, and I shall elaborate on this point at the end of the section. However, taken as it stands as an argument against any kind of axiomatization in evolutionary work, I think that it fails, for immediately one wants to ask when exactly a theory does achieve the 'tidiness and completeness' requisite for an axiomatization? But as soon as this question is asked, then one can see that Goudge's somewhat sweeping position is founded on a rather naive view of theory-construction. Goudge seems to imply that first of all a scientist gathers together a colossal body of information about his subject—concepts, laws, explanations, and such like—and only when he has done this does he then sit down and try to build an axiomatic system which incorporates all of this information. The real story is, of course, quite different. Theory building and the discovery of new ideas often go hand in hand. This truth applies even to axiomatic theory building, for in trying to axiomatize a limited amount of information, one often discovers ways of improving and extending one's information.

Of course, there is always a danger that one might become too committed to one axiomatic system too early; but, in fact, such a danger does not seem to be overwhelmingly great in the case of evolutionary theory. Indeed, if one studies the history of evolutionary theory, one finds that already evolutionists have demonstrated their preparedness to change their axiomatic systems if need be. Although the theory that Darwin gave in his *Origin of Species* is far from being stated with full deductive rigour, it seems clear that Darwin attempted to lay out the arguments at the heart of this theory in a fairly formal sort of way. In particular, what he did was argue first to a struggle for existence, and then from this, to natural selection. Both of these sub-arguments start with premises which are empirical generalizations and which Darwin clearly thought to be laws, and from the premises the conclusions are derived (with more or less rigour). Thus, for example, concerning the struggle for existence Darwin wrote:

A struggle for existence inevitably follows from the high rate at which all organic beings tend to increase. Every being, which during its natural lifetime produces several eggs or seeds, must suffer destruction during some period of its life, and during some season or occasional year, otherwise, on the principle of geometrical increase, its numbers would quickly become so inordinately great that no country could support the product. Hence, as more individuals are produced than can possibly survive, there must in every case be a struggle for existence, either one individual with another of the same species, or with the individuals of different species, or with the physical conditions of life. (Darwin, 1959 ed., 147)

Hence, for Darwin selection seems to have been something which must be proven within his system, immediately from the struggle for existence (together with the right kinds of organic variations), ultimately from such things as the potentially geometric population explosion. However, as we have seen from examples already discussed, in the modern theory, although evolutionists certainly try to give reasons why there is selection in particular cases, they do seem to take selection as something basic. For example, in the explanation of balanced polymorphism, as we saw, within the theory the existence of selection is given in the axioms rather than proven as a theorem. In other words, whatever other changes have occurred since Darwin wrote, there seems to have been a change of emphasis about which parts are to be presented formally, for it is certainly the case that the modern theory presents formally parts which Darwin did not. Darwin's views on heredity were a notorious hotch-potch of fiction and improvisation without any cohesion. Today, as we have seen, heredity is the subject of the most formal part of evolutionary thought. But this change from Darwin to the present would seem to end any doubts that Goudge has. Darwin's theory was axiomatic (albeit, only in parts and that very loosely) and so also is the modern theory (again, only in a sketched form); nevertheless, this has not prevented some very thorough developments within the theory. Indeed, through what has probably been the greatest development of the theory has come a much increased reliance on the axiomatic method. Population genetics as was sketched in the past two chapters is obviously a much more extensive and rigorous attempt at an axiomatic system than are the passages in the *Origin* (like that just given) which approximate to the axiomatic form. Hence, through the incorporation of population genetics into evolutionary theory there has been an extension of the axiomatic nature of evolutionary theory. Moreover, in conclusion one can add that even if axiomatization had retarded development, Goudge could only argue that an axiomatization for Darwin was a bad thing. His argument would not show an axiomatization to be a bad thing, either now or in the future. (A detailed analysis of Darwin's arguments is given in Ruse, 1971a.)

Finally, we come to Goudge's third argument, one which is extremely popular amongst those who would argue that evolutionary theory is different. It revolves around the supposedly peculiarly 'historical' nature of the theory of evolution. Goudge argues as follows:

The theory of evolution contains as an essential part of its structure a number of historical statements. These deal with such phenomena as phylogenetic trends, adaptive radiation, the extinction of past forms of life, etc. But no evolutionist would be likely to agree that the statements of natural history can be deduced as logical consequences of a set of postulates. Consequently, the presence of historical statements makes it impossible to construct a full axiomatization of the theory. (Goudge, 1961, 16)

Again I think that Goudge's argument fails. Clearly, one *could* deduce statements of natural history from a set of postulates. Indeed, if one were so inclined, one could deduce anything at all from a set of postulates, just so long as one used the right postulates. Presumably, what Goudge really means is that given only a set of (non-contradictory) postulates which do not make references to particular times, then one could not infer any conclusions which make references to particular times, and if one excludes trivial counter-instances (such as p, therefore p v r, where p is non-historical and

c

r is historical), this claim seems to be well-taken. However, even though this weakened claim I am ascribing to Goudge is true, and even though an axiomatized evolutionary theory might not contain references to particular times in its premises, the existence of evolutionary statements which do refer to particular times does not rule out the possibility of an axiomatized version of evolutionary theory. The reason for this is that if in fact the grounds given are sufficient to rule out the possibility of an axiomatized evolutionary theory, then we must also rule out the possibility of axiomatized physical theories, for in physics as well as in evolutionary biology, we find statements which are tied to particular times. This conclusion is obviously ridiculous, and as soon as we consider the relationship between physical theories and those statements about physical phenomena which refer to particular times, we can at once see how thin Goudge's arguments really are. What we find, for example, is that statements such as 'the ball hit the ground at time *t*' follow from the basic premises of Newtonian mechanics *together with* certain other statements. It is these latter statements, rather than Newtonian mechanics, which tell us about the ball's behaviour at other times (earlier or later). In other words, theories (of physics and chemistry) act a bit like sausage machines—you cannot get sausages out unless you put meat in, and similarly, you cannot get temporal statements (i.e. statements which refer to particular times) out unless you put temporal statements in. But, and this is the point that Goudge seems to have overlooked, you can put temporal statements in. Thus, from a physical theory (one which is axiomatized) together with these statements one can deduce further temporal statements.

There seems to be no reason why the same should not hold for an evolutionary theory, even one which is clearly axiomatized. Indeed, if we are to judge from what we have so far seen of the theory, very much the sort of relationship does seem to exist between the theory and statements referring to particular times. Consider, for a moment, the explanation of the evolution of Darwin's finches. Now admittedly, what we try to derive in a case such as this are statements referring to particular times—for instance, in this case there are statements about the distribution of the birds on the archipelago at this present time. Statements, for example, showing how there are at present more endemic forms of finches on the more isolated islands than there are on the central islands. However, although we derive statements like these from statements about selection, mutation, and the effects of isolation—statements which are clearly part and parcel of evolutionary theory and which make no reference at all to particular times—other statements are also involved, statements which refer to actual conditions in the present and immediate past.

For example, there is the statement that the trade wind blows from the south-east quarter, where lie the islands of Chatham, Hood and Charles. This, I assume, is a statement implicitly informing one that such a wind blows now, and has done since the birds started to evolve on the Galapagos.

However, if Goudge is right, then even a statement about trade winds has to be part of evolutionary theory, for it is from such statements that we derive other temporally bound statements—not from statements about the theory of population genetics. But this conclusion seems implausible. If anything, the statement about the trade wind belongs to meteorology. There is certainly nothing biological at all about it. In fact, even those temporally bound statements which do refer to organisms do not seem to be particularly *evolutionary*—unless by definition one makes every statement about organisms 'evolutionary'. Take, for example, Lack's claim that there were no other passerine birds when the finches first came to the Galapagos Archipelago. Why is this statement, *in itself*, any more evolutionary than the statement that some specified beam has only a five-pound weight attached, Newtonian? Obviously, we do not think the latter statement particularly Newtonian, although this is not to deny that we might be able to use it in conjunction with Newtonian mechanics to derive other statements. Similarly, there seems no reason why we should not treat the former statement about the non-existence of passerine birds on the Galapagos in the same way. It, and like statements, can be used in conjunction with evolutionary theory to derive other statements; but, in itself, such a statement is not part of the theory. In other words, there seems no reason to analyse evolutionary theory in a way different from the analysis needed for physical theories.

What has led Goudge to argue as he does? I suspect he has made a mistake which is fairly commonly made about evolutionary theory, in particular, he has confused the *theory* of evolution, the thing which tells one about the mechanisms of evolution, with descriptions of the *historical paths* taken by evolving groups of organisms ('phylogenies'). Phylogenies are obviously historical; but although we may come to learn about them and understand them through the theory of evolution, they are not themselves part of the theory which rather is, in this sense, no more historical than any other theory. One can, for example, similarly distinguish between the non-historical Newtonian mechanics and the historical descriptions of the paths planets took in the past. This is not to deny that because evolutionists spend comparatively more time studying the history of organisms than physicists spend studying the history of planets, evolutionary studies seem more historical than physics. But this does not in itself make the theory of evolution historical in the sense that it is tied to particular events in the Earth's history. (A different point, and one which I shall be considering in Chapter 10, is whether biological theories in general contain concepts with any kind of historical dimension. Here, I have not argued for or against this; but neither has Goudge. Its resolution has no relevance to the question of axiomatization.)

Finally in this chapter, let me pick up the point that I made about Goudge's argument against premature axiomatization having an element of truth. Goudge's arguments do not show that axiomatization *per se* is a bad thing; but as I said earlier I do share what I think is a common concern with Goudge that it not be thought that biologists do or ought to axiomatize evolutionary theory, *no matter what the cost*. I am not rigidly against axiomatization in principle or in practice as Goudge seems to be (and as biologists are not); but it is certainly the case that evolutionists do face enormous practical and theoretical problems, and there is a constant reshuffling and reinterpretation of facts which a blind drive for axiomatization would seem to preclude. The danger behind any attempt to argue for some sort of axiomatic vision of a theory is, as Goudge rightly points out, that the reader is liable to be left with a kind of 'snapshot' view of the theory, where all the elements are apparently 'frozen' into place. Nothing could be less true of evolutionary theory—like a living growing organism, new parts at all levels are constantly appearing and old parts being sloughed off. The point I made about the development of the theory since Darwin illustrates this fact, and in Chapter 10 I shall show how modern genetical studies affect and will affect the Mendelian theory presented in the last two chapters. My claims in this chapter about evolutionary theory are that I think the theory has an overall unified structure of the form I sketched earlier, that parts are axiomatized, and that there is no theoretical objection to an overall axiomatization. I certainly would not deny that in fact such an axiomatization is not and probably will never be completed in such a way as to incorporate every area of evolutionary studies as we now know them, although I would point out that the axiomatization of population genetics, a major innovation since the *Origin*, does point to a much increased use of the axiomatic method in evolutionary biology. Perhaps the reader might feel that my position stretches too far my use of the term 'ideal' when I claim, as I have done, that the hypothetico-deductive model is in a sense the ideal of evolutionary theory. I would disagree—for reasons given and more strongly for reasons to come—but I do not want my claims to conceal the diffuse way in which evolutionary theory is usually presented to us.

5

THE THEORY OF EVOLUTION

II: EXPLANATION

The question only partly answered in the previous chapter was that of why one should think a deductive model could have any relevance to the links between the various levels of evolutionary theory. As I explained in an earlier chapter, this question is just one of a set of general questions about evolutionary explanations, for, as we have seen, scientific explanations seem to be directed towards two different kinds of things. First there are explanations of individual facts; secondly there are explanations of laws. But, as we have also seen, there seem to be two basic questions about such explanations. First, is it the case that, as the covering-law model of scientific explanation stipulates, the thing explaining (the *explanans*) contains at least one law? This question seems to be directed essentially towards explanations of particular facts, since I doubt if anyone would deny that the explanation of laws (if they are explicable) calls for laws in the *explanans*. Secondly, there is the question of what kind of links exist between *explanans* and *explanandum* (thing explained). Is it the case that, as the covering-law model stipulates, the links must be deductive (or strongly inductive as some versions of the model allow)? Now this question applies both to explanations of individual facts and to explanations of laws. But the problem of the explanations of laws seems to be precisely that of the question of the relationship between statements at different levels of a theory, for how else does one explain laws like, say, Kepler's laws, other than by showing how they follow in some way from higher-level statements in the theory (like Newton's law of gravitation)? Thus it would seem that by addressing ourselves to general questions about evolutionary explanations, my claim that evolutionary theory is a hypothetico-deductive sketch will not be ignored.

No one would deny that most evolutionary explanations as actually given do not satisfy the covering-law model; but, one might want to ask, why should anyone deny that the model is in some important sense an ideal—the ultimate goal of evolutionists? I think one of the chief motivations of the critics of the model stems from the fact that, as has often been noted, closely related to the covering-law model of scientific explanation is a model for scientific prediction. In particular, if one has a covering-law explanation, then this could have been used for a prediction—the only difference being that in an explanation, the conclusion of one's argument (the *explanandum*) has already occurred, whereas in a prediction, the truth stated by the conclusion has yet to occur. (The converse does not seem to be entirely true, for although many scientific predictions are of the covering-law form, some are clearly not.) But the critics of the model, bearing in mind this close relationship between covering-law explanation and prediction, have taken note of the fact that although evolutionary theory seems fairly explanatory, as an organ of prediction it seems less than first class. Through the theory one might be able to throw some explanatory light on the evolution of certain groups of organisms, for example, Darwin's finches; yet one could never have predicted this evolution before the event. Of course, covering-law modellists can hardly deny this fact; but this does not shake their faith in the model. They find the source of the lack of predictive power of the theory partly in the fact that many of the explanations of the theory are only sketched in because vital premises and links are unknown and therefore omitted, and partly in the fact that it is only with hindsight that one can recognize the truth and importance of many of the facts needed for evolutionary explanations. (Who, for example, could have realized that the absence of passerine birds on the Galapagos would have been so important in the evolution of Darwin's finches?) However, the critics argue that the lack of predictive power of evolutionary theory points to the fact that the covering-law model is not appropriate for evolutionary theory. The theory, they argue, has perfectly adequate models of its own—models which explain, but which do not predict. In this chapter I shall consider three such models, all of which I shall show are less than satisfactory. I hope that my criticisms will reveal positive reasons for accepting the covering-law model in the context of evolutionary studies; but to conclude the chapter I shall explain why I find it intuitively implausible to suppose that evolutionary explanations have a model which is not that of the explanations of the physical sciences (whose model I take to be the covering-law model). (In this chapter I shall be arguing for the necessity of covering-law explanation, for it is against this that evolutionary theory has been used. I shall pay little attention to the question of whether all covering-law explanations are good scientific explanations, although it is clear that some are not. A deduction of a law from itself fits the model; but it is not much of an explanation.)

5.1 Characteristically genetic explanations

Gallie (1955) argues that within evolutionary biology we find good completed explanations which cannot in any sense be used for prediction. (Scriven, 1958, offers a very similar argument.) About such explanations, which he calls 'characteristically genetic explanations', Gallie writes as follows:

The first prerequisite of a characteristically genetic explanation is that we shall recognize the *explicandum* [i.e. *explanandum*] as a temporal whole whose structure either contains certain persistent factors or else shows a certain definite direction of change or development. Thereupon we look for an antecedent event, the *explicans* [i.e. *explanans*], which can be accounted a necessary condition of the explicandum, on ordinary inductive grounds (observations of analogous cases), but more specifically on the ground of a probable continuity—in respect of either persistent factors or of direction of change—between explicans and explicandum. (Gallie, 1955, 155)

Unlike the covering-law explanation, therefore, the characteristically genetic explanation makes no appeal to laws (at least, not in the *explicans*), nor does it attempt to specify the kinds of link between *explicans* and *explicandum* demanded by the covering-law model. A necessary condition is not necessarily grounds for a deductive inference, nor even for saying that the *explicans* gives strong inductive support to the *explicandum*. (Following Gallie, I shall in this section use the alternative '*explicans*' and '*explicandum*' for the more common '*explanans*' and '*explanandum*'.)

The example which Gallie chooses to illustrate his model of evolutionary explanation is one which occurs with rather monotonous frequency in philosophical discussions about biology—it is the explanation of the long necks of giraffes. He writes:

We have all been told that the species giraffe has arisen through a succession of longer and longer necked species because, at every stage in this sequence, possession of unusually long necks gave the species in question certain advantages, e.g. as regards food-getting, and thus enabled it to survive and leave descendants whose inherent variability allowed for further changes in the same general direction. But what is the force—or the legitimate interpretation—of the 'because' in this explanation? If it means simply that the giraffe and the intermediate species would probably not have survived unless they had possessed certain advantages connected with the possession of long necks, then it is an extremely plausible conjecture. There can be no doubt but that this is a conjecture which really explains: it enables us to pass from the relatively platitudinous assertion that the species giraffe could not have arisen except from the ancestry which, on grounds of continuity, it would appear to have had, to the assertion that unless its ancestry had displayed certain characteristics which are specifically mentioned, or unless this ancestry had conformed in certain specified ways to the general requirement of competition for survival, the species giraffe would not exist today. (Gallie, 1955, 157)

Now, let us look briefly first at Gallie's general model for evolutionary explanation, and then at the specific example which he offers us. As far as the first point is concerned, assuming not only that evolutionists want to explain but that they want to offer what are in some sense *good* explanations, then, as several commentators have pointed out, a model like Gallie's is altogether too weak. (See, for example, Montefiore, 1956.) The model allows as adequate explanations, 'explanations' which no good scientist would in any sense feel satisfactory. Consider, for example, the following candidate as an explanation: 'The explanation of the long necks of giraffes is that some 150 million years ago a group of small, warm-blooded, milk-giving animals (i.e. mammals) appeared, which, compared to their rivals (e.g. reptiles), had certain advantages (e.g. because they were warm-blooded they were not immobilised by cold).' This supposed explanation satisfies all the demands that Gallie's model makes of an explanation-candidate. The *explicandum* is obviously 'a temporal whole whose structure either contains certain persistent factors or else shows a certain definite direction of change or development'—indeed, it is the same *explicandum* that Gallie gives in his example. The *explicans*, animals with warm blood and with their consequent advantages, is an antecedent event which is as much a necessary condition of the *explicandum* as is ancestral giraffes with longer necks than their rivals (with their consequent advantages) —also, there is surely a continuity between the first mammals and today's long-necked giraffes. Nevertheless, one can hardly feel that the existence of the early mammals as such throws much in the way of explanatory light on the long necks of giraffes or on elephant's noses or human brains for that matter. (Of course, I realize that in my demand for 'good' explanations the critic will claim that I am begging a thousand questions; but I would suggest that one must have some standards, and my imaginary example does not fit them, whereas it does satisfy Gallie's model.)

It would seem, therefore, that before it can be counted as an adequate model of evolutionary explanation, Gallie's characteristically genetic explanation model must be strengthened in some way (or ways). Now, I do not at this point want to get bogged-down in too detailed a theoretical discussion; but it would seem initially that the strengthening of Gallie's model demands that in some sense one 'broadens' the necessary conditions called for in the *explicans*. Somehow, the existence of the early mammals, even given all of their favourable features, seems too far removed from the giraffe's necks —the first mammals are so far distant in time and evolutionary change that, necessary though they may have been, we feel that there is altogether too much of a (logical) gap for the mammals really to throw explanatory light on the giraffe's neck. Consider an analogous case. There is a rule that at the University of Guelph a student must take (and pass) thirty courses (of a particular kind) before he qualifies for a B.A. Suppose that we ask for an explanation of why a certain student deserves a degree, and we are told that he got a very high mark in introductory philosophy. As an explanation of why that student deserves a degree, there is too much of a logical gap if all we are told is that he passed one course, introductory philosophy, even though he did so with flying colours. The passing of introductory philosophy may indeed be a necessary antecedent condition of getting the B.A.; but, on its own, passing introductory philosophy is not enough to get the degree, or to explain why the student deserves the degree. Before we feel the meriting of the degree is explained, we must broaden the scope of the necessary conditions specified in the explicans. But what in fact will this broadening involve? Presumably in the case of the explanation of a student's qualifying for a degree, we will point out other necessary conditions he has fulfilled—namely, other courses he has taken and completed successfully. And similar sorts of things would seem to be called for in the above quasi-explanation of the giraffes' necks.

However, notice now what this addition of other (or more encompassing) necessary conditions entails. If we add further necessary conditions, that means that we get closer to a sufficient condition, for a conjunction of all the necessary conditions of something is a sufficient condition of that thing. A necessary condition of a thing x is something without which x could not occur; but if we have all the necessary conditions, then x must occur, that is, we have a sufficient condition of x. Thus, in the degree case, given that successful completion of 30 (specified) courses are all of the necessary conditions for getting a degree, if a student has successfully completed all of these courses, then given the rule about the sufficiency of 30 courses, no further demands can be made of him before he is granted his degree. He has achieved a sufficient condition for the degree, and to argue otherwise is to suggest there are other necessary conditions.

Similarly, if we had all of the necessary conditions of the giraffe's necks, we would have a sufficient condition of the necks.

The long and the short of this discussion seems to be that if characteristically genetic explanations are to be counted as *good* explanations, they must go at least some way towards being sufficiency-explanations, and that there is a minimum below which they cannot drop. (A 'sufficiency-explanation' is one whose *explicans* gives a sufficient condition for the *explicandum*.) Now, it might be claimed that this does not prove that for an adequate explanation one must go all the way to a sufficiency-explanation, nor even does it prove what a minimum allowable necessity-explanation might be (nor, how one might measure such a minimum—most cases are far more complex than where one has thirty, equally necessary, academic courses). Similarly, it does not prove that one's *explicans* must contain laws (although, I shall be taking up this question later in the chapter). However, I would suggest that it does seem that a (good) characteristically genetic explanation is starting to look more like a covering-law explanation, for if one demands a *deductive* link between *explicans* and *explicandum*, then one is demanding that the *explicans* state a sufficient condition for the *explicandum*. Even if one allows that an adequate covering-law explanation might specify only a strong inductive link between *explicans* and *explicandum*, then one is saying that the *explicans* must include reference to necessary conditions which, in some sense, conjointly approach a sufficient condition. Thus, whilst there might still be room for disagreement between supporters of the two models, characteristically genetic and covering-law, at least they are starting to talk in the same kind of language. With respect to the links required between *explicans* and *explicandum* they no longer seem so far apart as Gallie's discussion suggests. For both models, the sufficient conditions of the *explicanda* do at least seem relevant, and, to be honest, I would think that if the supporter of the characteristically genetic model still feels that the links between *explicans* and *explicandum* need not be those specified by the covering-law model, the onus is now upon him to show just what these links might be. In particular, he must show why, despite the fact that one must go a certain distance in the direction of a sufficient condition, even though it is (logically) possible to go farther, one reaches a point when it is just not necessary to go on. If he does not show this, and Gallie certainly does not, then in this respect the covering-law model seems an appropriate ideal. I shall pursue this problem no more here; but, as promised, at the end of the chapter I shall give a number of reasons why I think it intuitively implausible, at least in the case of evolutionary explanations, to suppose that cogent arguments could exist proving that the filling-out of a sufficient (*explicans*) condition could at some point become scientifically redundant. Of importance at this stage is the fact that Gallie has not supplied such a proof, and one is needed if his case is to hold.

Before going on, two points of clarification are needed. First, for a deductive link, one needs premises and rules of inference. In specifying sufficient conditions, one specifies only what would be in the premises in an argument—the conditions are 'sufficient' precisely because of the rules which have been given or are assumed. Thus, because and only because of the rule about 30 (specified) courses being sufficient, 30 courses are sufficient. Normally one does not have University senates laying down rules of inference and one has to rely on more general rules, like *modus ponens* (i.e. the rule of affirming the antecedent). I shall say no more about this point here; but from now on I shall assume that, unless special rules of inference are supplied, sufficient conditions are such that they, or more precisely, statements about them, can form the premises of deductively valid arguments which use no rules of inference other than those found in standard logic texts. This does mean that I can treat laws as conditions, necessary or sufficient; although I would point out that merely in allowing that a law can be a necessary condition, even though later in the chapter I shall be arguing that laws are necessary conditions of good scientific explanations, I do not thereby imply that the *explicans* of a good scientific argument must include a law. Readers who object to talk of laws as 'necessary conditions' may rest assured that no essential part of my analysis is dependent upon this terminology. Later I shall show that my case still carries if one refers only to particular things or phenomena by the term 'condition'.

Secondly, it must be admitted that my discussion of the link between necessary conditions and sufficient conditions has been simplified somewhat. I have argued that a sufficient condition must (in some sense) contain all of the necessary conditions, and from one point of view—a point of view which is adequate for this discussion —this is correct. However, there are things which complicate this analysis. Sometimes we have sufficient conditions which seem not to be or to contain necessary conditions. Consider the following case. Part of the regulations for getting a B.A. degree at the University of Guelph specifies that one of the required 30 courses must be a science course or equivalent. A sufficient condition for satisfying this science requirement (given the university rules) is the successful taking of philosophy of science. However, although the philosophy of science course is a sufficient condition, it is not a necessary condition. One could, for example, take geomorphology instead. Hence we seem to have here sufficient conditions which neither are nor contain necessary conditions (but where there do seem to be necessary conditions lurking somewhere in the background, for it is clear that one must do something to get the science requirement out of the way).

Obviously the cause of this situation, something my analysis has ignored, is the fact that sometimes we get a set of conditions (called 'contributory' conditions by Goudge), no one of which is on its own necessary, but one of which (i.e. any one of which) must be included in a sufficient condition. The necessary condition is in fact the *disjunct* of the particular set of contributory conditions. (One can, of course, have more than one set of contributory conditions.) Thus, in order to satisfy the science requirement it is necessary that one complete successfully *either* philosophy of science *or* geomorphology (or one of a number of other specified courses). Strictly speaking, therefore, a sufficient condition contains all of the necessary conditions or, where some of the necessary conditions are disjuncts of sets of contributory conditions, at least one contributory condition from each set. As I have said, this more accurate analysis of the relationship between necessary and sufficient conditions makes no real difference to my discussion of Gallie. The important thing about Gallie's model of explanation is that it does not require one to specify a sufficient (or near sufficient) condition—just some necessary conditions. My reply is that such a model is too weak. Neither Gallie's model nor my reply is much altered if one recognizes that some necessary conditions are disjuncts of contributory conditions. Taking note of these, Gallie's model now requires the specification of some necessary conditions and/or some contributory conditions (from different sets) but does not require that one have a sufficient (or near sufficient) condition. My reply is that one must approach a sufficient condition through the specification of necessary conditions (i.e. those which are not disjuncts of contributory conditions) and (where relevant) the specification of at least one member of each set of contributory conditions. (As Gibson, 1960, 187n, points out, one can certainly extend Gallie's model legitimately in this way, for in fact Gallie does sometimes give examples of conditions which would usually be regarded as contributory conditions rather than as necessary conditions.)

However, before leaving this point, there is one final thing which must be said. I admitted at the beginning of this chapter that I am concerning myself solely with the question of whether explanations in the biological sciences must at least have the covering-law model as an ideal in some sense, and I conceded that although this requirement may be a necessary condition of good scientific explanation, it is not a sufficient condition. Beckner (1967) has pointed out that in the quest for a model of scientific explanation which is sufficient as well as necessary, one must take note of which contributory conditions are being offered in a sufficiency-explanation. Not every contributory condition will do if in fact other contributory conditions (of the same set) have also been satisfied. For example, sufficient conditions for the avoidance of a small-pox epidemic (certain pertinent laws being assumed) are either the absence of the small-pox virus in the first place or an effective immunization programme. But we would probably not consider it much of an explanation of the avoidance of the epidemic if we were told merely about an immunization programme, if later we learned that there was no virus in the first place. I am sure that one can find a model of explanation such that difficulties of this nature can be avoided. But it is clear that the search for a fully adequate model of scientific explanation is much more complex than might appear at first sight.

Briefly, now, let us turn to look at the actual example which Gallie offers. I must confess that I have trouble seeing this example as offering strong support for the characteristically genetic model. Apart from anything else, the *explicans* contains something which strikes me as being remarkably lawlike. (I have admitted that understanding 'sufficient condition' in the way I specified in the last paragraphs, it does follow that necessary conditions could be laws as well as particular matters of fact. However, I do not think that Gallie intends any of his necessary conditions to be laws.) Consider the statement that 'unless this ancestry of giraffes had conformed in certain specified ways to the general requirement of competition for survival, the species giraffe would not exist today' (Gallie, 1955, 157). Frankly, unless one goes all the way with Smart and denies that biology has any laws at all, I fail to see how the demand that organisms conform to 'general requirements' can be other than something based on an assumption of nomic necessity. No one ever saw the ancestral giraffes; but still we feel able to say that certain things *must* have held for them, and this, after all, is what laws are all about.

However, admittedly, even if we grant this point, and I think in fairness Gallie would grant it—he seems not so much to deny the importance of laws as to deny that one would bother to put them in one's *explicans* (one would be keeping them up one's sleeve to justify the necessity of one's *explicans*-conditions)—Gallie is still certainly right in claiming that the 'explanation' is far from being something which offers us a sufficiency condition. But again, I run into trouble with Gallie's analysis. Missing from the explanation, as Gallie himself points out, is any mention of the genetics of giraffes, and, of course, without such mention one cannot go deductively from the longer-necked successful ancestors, to the present long-necked giraffes. Apart from anything else, long-neckedness might be something like sun-tan, which does not build-up over the generations.

But, if we accept Gallie's analysis, then the evolutionist's thinking on the question of giraffe-necks seems very odd. He (the evolutionist) knows something about genetics (specifically about giraffe-genetics), and indeed, after giving an explanation along the lines described by Gallie, he could and would then go on to offer explanations which utilize this genetical knowledge. (To deny this fact is to reveal that one has never looked at anything where evolutionists write for themselves—like the journal *Evolution*. Indeed, to deny this fact is to reveal that one does not know of Darwin's revisions in later editions of his *Origin* as he tried desperately to plug gaping holes which his ignorance about the laws of heredity left in the first edition, and it is also to deny the rationality of twentieth-century evolutionists' obsession with Mendelian genetics.) The evolutionist certainly could not fill out all of our gaps about the evolution of giraffes; but he certainly could (and would) show how genes for long-neckness would be transmitted from one generation to the next and how they could be passed through the whole population. However, if what Gallie writes were correct, we would then have two, quite separate, perfectly adequate, evolutionary explanations of the giraffes' necks. The one explanation is the one Gallie gives (i.e. an explanation making no reference to genetics), and the other explanation is Gallie's explanation plus genetical information. To me, the obvious way to interpret this kind of situation is to claim that the former explanation is a precursor of the latter—in other words, the latter is an improved version of the former. If it is not, then I cannot see why the evolutionist would even bother to provide the second explanation (the one with genetics).

Gallie, however, has to argue that the former explanation is perfectly adequate, in no need of supplement, even though he could not deny that the evolutionist then goes on to offer an amplified explanation. Admittedly, at one point Gallie says, rather revealingly, that 'characteristically genetic explanations do explain—do establish conclusions which it is very important to know, if only because they commonly provide the premises—the instantial or historical premises—of further explanations of predictive pattern' (Gallie, 1955, 156). Nevertheless, the major theme of his paper is that 'explanations in terms of temporally prior necessary conditions are commonly put forward when there is no good ground for accepting —and when indeed there is no consideration of—further explanation of a more complete, and in particular of a predictive, character' (Gallie, 1955, 157). If what Gallie writes were true, then it seems that, in some sense, if the evolutionist appeals to genetics to help in his explanation of giraffe-necks, he is wasting his time. He already has an explanation which is in its 'own way perfectly satisfactory'

(Gallie, 1955, 152). For myself, I cannot see this way as being other than as a first stage on the road to an explanation which appeals to genetics. Otherwise, there seems no continuity in what the evolutionist does—indeed, much of what he does seems unnecessary. (In fact, Gallie seems even to deny that the evolutionist can appeal to genetics, let alone that he would.)

I shall say no more directly about Gallie's model of explanation. Rather, I shall now turn to two somewhat different candidates for models of evolutionary explanation. Both of these models have been offered by Goudge, who, like Gallie, wants to replace the covering-law model as a guide to evolutionary explanations. In particular, Goudge offers models where there are appeals to sufficient conditions (like the covering-law model but unlike Gallie), but where there are no appeals to laws (unlike the covering-law model and like Gallie in that no laws appear in the *explicans*, but unlike Gallie in that no appeal to laws seems in any way possible). Let us take Goudge's two models in turn.

5.2   Integrating explanations

Goudge argues that at one level in evolutionary theory we find explanations of a wide range of phenomena, all of which are integrated by the fact that they share the same *explanans*. This *explanans* is a general historical statement about the broad outlines of life on this earth, such as 'Living organisms are all related to each other and have arisen from a unified and simple ancestry by a long sequence of divergence, differentiation and complication from that ancestry' (Goudge, 1961, 65). Goudge argues that a statement like this gives us a sufficient condition of the *explanandum* of an integrating explanation; but that there is no appeal to laws (in any way) in such explanations.

One of the examples of an integrating explanation offered by Goudge is of the presence of vestigal organs (like, for example, man's appendix). Goudge claims that such organs are explained by an appeal to a statement about the history of life on earth, such as that just given. His reconstruction of the explanation is as follows:

Most organs and structures observed in present-day animals have an adaptive function in relation to their way of life. By virtue of the uniformitarian principle, we infer that most organs and structures of past animals were likewise adaptive. Yet among present-day animals (and plants) vestiges are observed. Now the presence of these vestiges would be accounted for if they were inherited remains of organs and structures which once had an adaptive function in the lives of ancestral organisms. But we have plenty of evidence that such ancestral organisms existed; and it is reasonable to suppose that their adaptive needs, being different from those of their descendants, would require fully developed organs and structures of the sort now represented by vestiges. The presence of these vestiges is to be expected, given the general character of the history of life. Hence they are satisfactorily explained in terms of that history, together with certain assumptions which are plausible. (Goudge, 1961, 67)

I shall not at this point attempt a general analysis of Goudge's model. The main reason for this is that I am not quite sure how seriously Goudge would want us to take the model, for as we shall see immediately, Goudge's example differs drastically from the guide laid down by the model. In any case, most of the points I would want to make will occur and be discussed when I come to consider Goudge's other model. Here, I shall merely raise three points about Goudge's example, suggesting that if the example is typical, then the model leaves something to be desired (and that which is to be desired points in the direction of the covering-law model).

The first thing to be noted about Goudge's supposed example of an integrating explanation is that in the *explanans* we certainly have a great deal more than a general statement about the history of life. In particular, these 'certain assumptions which are plausible' read some highly theory-laden interpretations into the history, for to talk (today) of characters being 'adaptive' is to carry one right to the heart of evolutionary theory. I am not, of course, denying that one can legitimately make reference to evolutionary theory in this way; but, to do so takes one from a mere description of the history of life. Moreover, if we do refer to evolutionary theory, then we refer to a body of knowledge made up of laws. Hence, the *explanans* Goudge offers refers implicitly to knowledge of laws.

Secondly, the *explanans* does not give us a sufficient condition of the *explanandum*. Because character C was adaptive in the past, it does not follow that if present conditions are such that C would no longer be adaptive, I must now have character c, a vestige of C. For many reasons, all traces of C might have been eliminated entirely. Hence, here again, Goudge's example does not fit his model.

Thirdly, as soon as we try to fill in the gaps left in the *explanans* in order to make the example fit the model (i.e. by providing a genuine sufficient condition), we find that we must refer to genetics and to its laws. For instance, we might want to show how some vestiges are pleiotropically linked to other, still-adaptive characters, and that hence, overall, the genes for the vestiges enjoy a selective advantage. But, if we do something like this, then the explanation of vestiges starts to look like support for the covering-law model and for my conception of evolutionary theory, rather than like support for Goudge's integrating explanation model.

Now, there are a number of replies that Goudge might want to make to my criticisms. In particular, he might want to weaken the demand for a sufficient condition since, not very consistently, at one point he does talk of integrating explanations 'stating either a general sufficient condition or an important necessary condition of the phenomenon to be explained' (Goudge, 1961, 69). Since these possible replies by Goudge are probably the same as he would want to make to my criticisms of his second model of evolutionary explanation, I shall turn at once to look at this.

5.3   Narrative explanations

After they have made integrating explanations, Goudge argues that evolutionists turn next to work at a rather different level, wherein they formulate what he calls 'narrative' explanations. These are the kinds of explanation which purport to show why particular evolutionary events have occurred, such as the evolution of certain major classes of organism and the extinction of other classes. Once again, Goudge thinks that we find no appeal to laws, and that hence the right model of explanation is not the covering-law model. Moreover, Goudge explicitly gives the reason why he thinks there can be no laws—a reason which I think, together with evolutionary theory's low predictive power, lies behind a great many of the alternatives to the covering-law model. This reason is that the phenomena being explained by the evolutionist are *unique*.

Goudge writes that since the aim is to make major evolutionary

. . . events intelligible as unique, non-recurrent phenomena, recourse must be had to historical or 'narrative' explanations. The situation does not permit of being treated systematically in terms of general laws. Hence, narrative explanations enter into evolutionary theory at points where singular events of major importance for the history of life are being discussed. (Goudge, 1961, 70–1)

As an example of such a narrative explanation, Goudge quotes from an account by Romer which explains why vertebrates left the water and evolved into land creatures. Romer argues that the reason for this evolutionary step was, paradoxically, to enable amphibians to remain in the water. Their evolution into land animals was accidental.

'The Devonian, the period in which the amphibians originated, was a time of seasonal droughts. At times the streams would cease to flow . . . If the water dried up altogether and did not soon return . . . the amphibian, with his newly-developed land limbs, could crawl out of the shrunken pool, walk up or down the stream bed or overland and reach another pool where he might take up his aquatic existence again. Land limbs were developed to reach the water, not to leave it.

'Once this development of limbs had taken place, however, it is not hard to imagine how true land life eventually resulted. Instead of immediately taking to the water again, the amphibian might learn to linger about the drying pools and devour stranded fish. Insects were already present and would afford the beginnings of a diet for a land form. Later, plants were taken up as a source of food supply. . . . Finally, through these various developments, a land fauna would have been established.' (Goudge, 1971, 71, quoting Romer, 1941, 47–8)

Goudge suggests that this argument by Romer is typical of a kind of reasoning to be found within evolutionary theory. The important thing to notice is that no attempt has been made to deduce the thing which is being explained, the evolution of vertebrates living on dry land, from a law or number of laws. Rather, the event being explained has been *fitted into context* by showing how a number of events lead up to it. It has ceased to be isolated and thus falls into place. Goudge adds that we can consider the reasoning as something which tries to give us a *sufficient* condition of the thing being explained. Now, Goudge points out that if C is a sufficient condition for an event E, then C must contain all the necessary conditions for E, n_1, n_2, n_3, . . . and at least one of the contingent contributory conditions, $c_1, c_2, c_3, \ldots$ (As I have said, by 'contributory' conditions Goudge means a set of conditions which disjointly make a necessary condition—he fails to notice that one might have different sets of contributory conditions disjoining to make different necessary conditions. I shall continue to talk, as I have done, only of necessary conditions, including in these contributory conditions.) Thus Goudge claims that in a narrative explanation we get necessary conditions (e.g. that the amphibians must have found food on dry land) and contingent contributory conditions (e.g. that the amphibians walked overland). However, adds Goudge, we rarely if ever get all the necessary conditions. Only those which are most important or least expected are given—there are bound to be some conditions which are just not known. Hence, to a great extent, narrative explanations will necessarily be conjectural—such explanations will often just be *possible* explanations and will make great use of such terms as 'could', 'would' and 'might'.

Because of this (as well as because of the absence of laws) Goudge argues that narrative explanations will never fit the covering-law model. He writes:

It may be tempting to try to make this pattern conform to certain familiar 'models' of explanation. The first is the model of a hierarchical deductive system, often declared to be the ideal form in which the theoretical part of every science ought to be cast, or, at any rate, to which the theoretical part of every science ought to approximate. The attraction exerted by this model is mainly due to its successful use in connection with physics. But it does not seem to be applicable to patterns of historical explanation in biology. Certainly, the temporal sequence of events specified by the above example is not such as to permit each event to be deductively inferred from its predecessors. The presence of many contingent, contributory conditions in the sequence makes it idle to hope that the statements constituting the explanatory pattern can be organized deductively, or even, perhaps, axiomatically. The deductive model is, therefore, the wrong one to have in mind at this point, however relevant it may be elsewhere. (Goudge, 1961, 74–5)

Narrative explanations, consequently, tell a 'likely story'. They cannot be forced into the deductive model, and they do not appeal to laws.

Goudge concludes his discussion by considering a number of possible objections to his model, the most important of which is the following. Suppose one has a narrative explanation of some particular evolutionary phenomenon E, and that one has a possible sufficient condition for E, namely s. Surely it could be argued that one does have a law here, namely that whenever a condition of the type s obtains, there will be a phenomenon of type E. Further, one can now set up the explanation in deductive form with s and this law as premises, and E as the conclusion. Hence, narrative explanations can always be converted into a kind which fit the deductive model.

Goudge's reply to this objection is as follows:

Whenever a narrative explanation of an event in evolution is called for, the event is not an instance of a kind, but is a singular occurrence, something which has happened just once and which cannot recur. It is, therefore, not material for any generalization or law. The same is true of its proposed explanation. What we seek to formulate is a temporal sequence of conditions which, taken as a whole, constitute a unique sufficient condition of *that* event. This sequence will likewise never recur, though various elements of it may. When, therefore, we affirm 'E because s', under the above circumstances, we are not committed to the empirical generalization (or law) 'Whenever s then E'. What we are committed to, of course, is the *logical principle* 'If s then E', for its acceptance is required in order to argue 'E because s'. But the logical principle does not function as a premiss in an argument; the affirmation, 'E because s', is not deducible from it. (Goudge, 1961, 77)

In other words, plausible though this objection may seem, Goudge argues that it does not in fact constitute a damning counter-example to his thesis.

Let us examine critically this second model that Goudge offers us. Unfortunately, as in the case of integrating explanation, since Goudge's example differs so much from the guide-lines laid down by his model, I find it difficult to try to criticize the model in abstract. I shall therefore turn back to Goudge's example of an event supposedly calling for a narrative explanation, the evolution of land-vertebrates, and I shall see whether Romer's explanation seems better understood in terms of the narrative or covering-law model, remembering that Goudge gives two reasons for preferring the narrative explanation model. The first reason being that explanations such as Romer's supposedly appeal to no laws, and the second being that evolutionists rely too heavily on terms like 'could' and 'would' for their explanations to fit the covering-law model.

Now, it should be noted immediately that, taken completely *in isolation*, Romer's account of the evolution of the land-vertebrates falls short of being an adequate explanation, considered either from the covering-law point of view or from the narrative viewpoint. It is easy to see that this is the case as soon as we remember that both models demand sufficient (or nearly sufficient) conditions. Nothing like a sufficient condition for the evolution of land-vertebrates is offered by Romer. The conditions that Romer gives are that there was a drought and that (initially) the amphibians had to have water. If these jointly constituted a sufficient condition (for the evolution of land-vertebrates or even the beginning of such an evolution), then it would have been impossible for anything to have happened other than that the amphibians began to develop land limbs, because this is precisely what is meant by calling such conditions 'sufficient'. As we have seen, to say that A is a sufficient condition for B is to say that given A, B must follow. However, not only was the development of limbs not the only response which *might* have been made to droughts by organisms like the amphibians (i.e. organisms which needed water), it was not the only response which such organisms *in fact* made to droughts. Another group of organisms, lung-fish, when faced with a situation similar to the amphibians, developed the ability to remain during droughts in a dehydrated state of suspended animation baked in the mud at the bottom of dried-up pools. When the rains came, they were reactivated and continued their normal (fishy) existence. Furthermore, they have been doing this for millions of years now. Why did not the amphibians evolve in a similar manner, rather than going the way that they actually did? Until this question is answered we will not have a sufficient condition for the actual evolutionary event which did occur, and thus, considered with respect to either of the two models, we will not have a totally adequate explanation. That this point is well taken is, I think, indicated by the fact that even Goudge is prepared to admit that the explanation as it stands is not complete, because as we saw, he himself points out that several important factors, vital for the occurrence of the phenomenon being explained, go unmentioned. For example, one necessary but unmentioned condition was 'the existence of a group of fishes functionally able to move on land and genetically capable of improving this adaptation' (Goudge, 1961, 73).

Let us look in a little more detail at some of the questions which

Romer's 'explanation' invokes—questions which suggest that despite the conditions which Romer gives, the evolutionary path of the vertebrates might have been other than that which it actually was. Consider the following three. Why should the amphibians be able physically to develop limbs at all? Why, at first, was it so important for the amphibians to stay in water? Why, later, even though there was food on land, did the amphibians not remain far more tied to their pools than they did? Unless some of these questions are answered, then one cannot claim to have given a sufficient condition for the evolution of land-vertebrates. However, and here we get a crucial objection to narrative explanation, I fail to see how any of the questions posed can be answered without making some reference to laws. Consider the first question, why amphibians should be able to develop at all. To answer this, even briefly, one must necessarily sketch some of the principles of genetics (particularly population genetics). One must explain the nature of mutations, how these affect the physical characteristics of organisms, how these can be passed on from one generation to the next, and how such new characteristics can spread throughout a group. Now, obviously one's answer is not going to be based upon one's experience of evolving amphibians—rather one is going to appeal to the knowledge that one has of such matters as they apply in contemporary organisms, and one is going to suggest that it is reasonable to accept that something similar held for the amphibians. For example, one will suppose that the reason why amphibians had the capacity to develop land-limbs had nothing at all to do with the needs of the amphibians—one's justification for this supposition being that mutations today occur 'randomly' (i.e. without respect for the needs of their possessors). Suppositions of this nature are clearly lawlike, since they are generalizations which one believes hold even in unexamined instances. That is, they are generalizations which are held to be empirically or nomically necessary. Hence, it would seem that one must appeal to laws to answer the question of why the amphibians were able to develop limbs, and until this question is answered, we do not have a sufficient condition for the evolution of land-vertebrates. Actually, in a case like this, we seem to have a dual appeal to laws. First, we must use laws to infer that the organisms involved would have genetical systems of a type shared by contemporary organisms. Although laws would be needed here, I do not think that by virtue of this use they would go into one's explanation. But secondly, one would then need to appeal to laws in one's explanation, because specification of the genetic systems alone would not constitute a sufficient condition—in the sense of 'sufficient' understood here—for the evolutionary change of the organisms. Given the systems alone, a certain kind of change does not *have* to follow, unless one also assumes laws—but 'having to follow' is what sufficiency is all about.

Consider also the kind of answers which should be given to the second and third questions. One must show why the amphibians were so keen at first to stay in the water, and why, later, they freed themselves from the water as much as they did. Answers to both questions will revolve around the notion of 'adapative advantage', showing that what actually happened was adaptively preferable to any other course of action in the circumstances. However, as we have seen, to talk knowledgeably of adaptive advantage presupposes an acquaintance with one of the key concepts of evolutionary theory, namely natural selection, since to say that something has an adaptive advantage is to say that it will be favoured by selection. But, as we have also seen, modern evolutionary thought locates natural selection within that law-network known as 'population genetics'. Hence, it is difficult to see how answers to either of the two questions could avoid making fairly substantial references to basic biological laws.

Without going into further details, it must by now be clear that all three of the above questions can be answered only by appealing to various laws of nature. In other words, it would seem that if Romer's explanation is to be considered adequate, even with respect to Goudge's model of narrative explanation, we must assume that there is an implicit reference to law. If we do not assume this, then we must allow that we do not have a sufficient condition (even an implicit one) for the evolution of the land-vertebrates and hence we do not have a narrative explanation—something which Goudge claims we do have. However, if once it is conceded that there is a reference to law, then (from this point of view) there seems no reason to prefer the narrative explanation model over the covering-law explanation model. One can say that what Romer gives us is a covering-law sketch—the laws are there, albeit hidden, and to assume that they are not presupposed is akin to assuming that, confronted with a slightly theory-laden description of an eclipse, there are no references to laws in physics. The physicist does not head each page with Newton's laws, no more should the biologist need to head each page with laws from Darwin and Mendel, although it is not without relevance that the most important book on evolution by a paleontologist since the war, Simpson's *The Major Features of Evolution*, starts with five chapters on genetics, and indeed the author notes that his aim is 'to try to apply population genetics to interpretation of the fossil record and conversely to check the broader validity of genetical theory and to extend its field by means of the fossil record' (Simpson, 1953, *ix*). Nor is it without relevance that Romer's discussion comes in a book designed for the general (i.e. non-biological) public.[1]

There is one objection that Goudge might want to make at this point. Although he himself makes little effort to unpack the notion of a 'sufficient condition', he might nevertheless complain that common usage implies that a sufficient condition, in the absence of any special rule of inference, is something which *together with laws* can serve as the premises of a deductive argument, rather than, as I am using the term, as something which *on its own* can serve as the premise of a deductive argument (the difference being that I include the laws as part of the condition). Thus, Goudge might argue, the specification of a sufficient condition as such does not require an appeal to laws. Hence, in a case like the evolution of land-vertebrates, what we need is the specification of the genetics of the organisms involved and we need not add extra premises containing laws of genetics. However, I do not think that this move helps Goudge very much, since there is an obvious reply. If one adopts Goudge's usage of the term, then one must still appeal to laws in order to know that one has in fact got a sufficient condition (just as I think Gallie admits that one needs laws to know one has a necessary condition). Presumably one's explanation now consists of the specification of a particular condition as *explanans*, and, if one is determined not to have laws in the *explanans*, one must use a rule to the effect that if one has a condition of a type mentioned in the *explanans* then, by virtue of certain laws, other conditions will obtain—these other conditions forming the *explanandum*. Without using such a rule (i.e. by using instead a rule such as *modus ponens*), there is no reason to suppose that the *explanans* is in any way relevant to the *explanandum*, let alone deductively relevant, which, by virtue of his talk of sufficient conditions, I assume Goudge thinks an *explanans* should be (although it must be confessed that he is not always consistent about this). But now, whilst admittedly we do not have any reference to laws in the premises, we do have explanation-arguments which are formally equivalent to covering-law arguments, inasmuch as they both make essential references to laws. Hence, Goudge is back where he started. (But see Alexander, 1958, for a discussion of why it is better to treat laws as premises rather than as rules of inference or as things justifying such rules.)

[1] In passing, it is perhaps worth noticing that even Goudge himself admits that there is some justice in what I am arguing, for he does concede that 'The boundary of background conditions of a particular narrative explanation undoubtedly contain a vast array of prior information, assumptions, inductive generalizations, etc.' (Goudge, 1961, 76). However, he then goes on to quote Popper to the effect that 'If laws are involved, they are so trivial "that we need not mention them and rarely even notice them"' (Goudge, 1961, 76). If my points are well taken, then the laws involved are far from trivial, and we would do well to notice them.

The next counter-move that Goudge might make is to suggest that one can have a sufficient condition in his sense; but still not have to use a rule incorporating (or justified by) laws. In fact, this is a move that Goudge makes, but before turning to it, let us first consider Goudge's second reason for rejecting the covering-law model in favour of the narrative model. (In discussing this reason we can consider 'sufficient condition' in either Goudge's sense or mine.) Goudge's reason, it will be recalled, revolves around the impossibility of finding enough necessary and contributory conditions to make up a sufficient condition of a phenomenon being explained. Because of this he argues that rarely, if ever, will it be possible to find all the necessary and contributory conditions of the phenomenon, that biologists end up with *possible* explanations making great use of such terms as 'could', 'would' and 'might', and that 'the deductive model is, therefore, the wrong one to have in mind at this point' (Goudge, 1961, 74–5).

Even if we interpret Goudge as sympathetically as possible, it is difficult to see how this criticism can fail to rebound on narrative explanation any less than on covering-law explanation. There are perhaps three ways in which one might take what Goudge writes. First, Goudge might merely be pointing to the fact that often biologists encounter so many unknowns that even if they wanted to, they could not supply all the required necessary and contributory conditions—all they can do, as it were, is block out the main conditions, so one can see the outlines of a sufficient condition. Now this claim seems to me to be true—in *Darwin's Finches* Lack is forever admitting that he has to leave phenomena unexplained—but I fail to see how the claim supports Goudge (and, more importantly, condemns the covering-law model). Narrative explanation, just as much as covering-law explanation (more so than some versions of the covering-law model), requires that a sufficient condition be given —if it is impossible to give one, then we can have only a sketch. However, presumably this conclusion holds whether one supports covering-law or narrative explanation. Hence, there is no special reason here for preferring the latter kind of explanation to the former.

The second thing that Goudge might mean is that often evolutionists can only put forward tentative conditions, that is, conditions which might possibly be true, but which it would be rash to claim definitely are true. Here it would indeed be a bit naive to claim that we even had a sketch of a covering-law explanation—as far as that model is concerned we would have no more than a very vague proposal. However again, although I agree that this is often the case, I cannot see that Goudge's argument supports his position. If we do not know which are the true conditions, then I fail to see how we can claim to have a narrative explanation either. Surely he would not want to claim that for narrative explanation (as opposed to covering-law explanation) the truth or falsity of the premises is quite immaterial? If so, I hope no biologist ever hears him say it.

The third thing that Goudge might mean is something rather more drastic. He might be saying that occasionally (indeed perhaps often) one can provide only a few of the required necessary and contributory conditions and that these fall far short of a sufficient condition. He might be saying further that although this could never be considered a covering-law type explanation, in such situations narrative explanation can loosen its requirements and provide a perfectly adequate explanation form, even though a sufficient condition is not even approximated. Although this interpretation does conflict with what Goudge has said earlier about narrative explanations appealing to sufficient conditions, it would perhaps give him a way of drawing a strong line between the two models. Unfortunately, I would suggest that even if Goudge were to preserve the autonomy of narrative explanation in this way, it is only done so at a very heavy price, because then we are right back to a model of explanation like Gallie's, and as we have seen, such a model is too weak. Hence, it would seem that this interpretation of Goudge, like the other two interpretations, fails to point favourably in the direction of his narrative model of explanation. The covering-law model seems just as adequate (and the same applies if Goudge tries to defend his integrating-explanation model by a similar move).

To conclude this look at narrative explanation, let us look now at the hypothetical objection that Goudge advances against his position and his reply to it. (This reply would also be Goudge's counter-move to my suggestion that even if one takes laws out of one's premises, one must appeal to laws in one's rules of inference.) The objection to Goudge's position supposes that if one supplies a (Goudge-like) sufficient condition s for an event E, then one is thereby justified in claiming to have the law, 'Whenever s, then E'. Moreover, it is claimed, this law can function as a premise in a deductive explanation of E. Goudge's reply is that s and E are unique and unrepeatable, and hence cannot form parts of a law. 'Whenever a narrative explanation of an event in evolution is called for, the event is not an instance of a kind, but is a singular occurrence, something which has happened just once and which cannot recur. It is therefore not material for any generalization or law.' As I pointed out earlier, this is a reason which lies behind many challenges to the covering-law model (that is, challenges based on evolutionary theory). Thus, Goudge's argument deserves special care.

Now, let me concede immediately that if events *s* and *E* are in fact entirely unique and unrepeatable, then Goudge's claim that the evolutionist is not committed to a lawlike statement is well taken. Consider for a moment the law-*schema* 'Whenever an instance of *A* occurs, an instance of *B* will also occur'. As we saw in Chapter 2, what this is asserting is that there is some kind of necessity between the occurrence of instances of *A* and the occurrence of instances of *B*. Further, although it may indeed be the case that instances of *A* *have* occurred very seldom, if indeed they ever occurred at all, it does (and must) leave open the possibility that an instance of *A* *could* occur frequently, not only in the past, but in the present and future also. Take for example the law 'On any celestial body that has the same radius as the earth but twice its mass, free fall from rest conforms to the formula $s = 32t^2$.' This may never have been satisfied and never be satisfied; yet because it is a law, there is no logical bar to its being repeatedly satisfied. If for some reason *A* were absolutely one of its kind, then there could be no law of the form 'Whenever we have an instance of *A*, we get an instance of *B*'. But, having said this, what I would dispute is Goudge's claim that evolutionary events such as we are considering are one of a kind in the required sense. Goudge's claim is that they are 'unique'; but I suggest that further investigation reveals an ambiguity in the term 'unique' and that when this ambiguity is exposed, Goudge's position is seen to be based on sand. In other words, the statement 'Whenever *s*, *E*' is a law (or, as I would prefer to put it, the statement 'Whenever we have an instance of *s*, we have an instance of *E*' is a law).

Consider for a moment some senses of the term 'unique' and let us see how these relate to the problem of whether or not an event like the evolution of land-vertebrates is unique in such a way that its conditions cannot be put into a law. (My discussion at this point is heavily indebted to the analysis of 'unique' in Gruner, 1969.) One thing one might mean if one says that an event or phenomenon *E* is 'unique' is that it shares its spatio-temporal coordinates with nothing else. The (Goudge-like) sufficient condition for the evolution of land-vertebrates is certainly unique in this sense; however, this kind of uniqueness is no bar to the application of law, because everything else is unique in this sense. Any simple pendulum has its own unique spatio-temporal coordinates; but this uniqueness does not stop us from asserting the law 'Whenever one has a simple pendulum, the period is proportional to the square root of the length.' Every law abstracts to the extent that it ignores, as it must, spatio-temporal coordinates. Consequently, although instances of type *s* and *E* have their own unique coordinates, this in itself is no barrier to their being part of a law. Another thing one might mean by 'unique' is that a thing has nothing at all in common with anything else. Here obviously one could not embody it within a law. However, the conditions leading to the evolution of land-vertebrates are not unique in this sense. There have been many droughts, and as I pointed out, there have even been other organisms which had to evolve to survive such droughts. We must therefore turn to yet another sense of 'unique', if we are to elucidate the special sense in which the conditions for the evolution of land-vertebrates are unique. I think it is fairly clear that the sense intended here is that never before (or since) have we had such a particular *combination* of conditions—the droughts, the amphibians, the genetic potential, the food on land, and so on. In this sense, it must be agreed that the conditions and the event which we are considering are unique. The question which needs answering therefore is whether this kind of uniqueness is a bar to the application of law. I cannot see that it is. Logically speaking, it would seem that the only things which cannot be repeated about such a combination is the particular conjunction of space and time; but as we have just seen, this type of uniqueness is not troublesome. All the other conditions could recur in the same conjunction many times, either on this planet, or perhaps (with some good possibility) elsewhere in the universe. That such conditions have not as a matter of fact been combined more than once does not stop them from being in a law, any more than that as a matter of fact we have probably *never* had a combination of molecules making a celestial body with the same radius as the earth but twice its mass stop us from (properly) calling the statement that 'on such a body free fall from rest conforms to $s = 32t^2$' a law. To argue otherwise would seem to require some sort of ontological argument in reverse (proving the logically necessary non-existence of repetitions of major evolutionary events). Since I doubt that such an argument can ever be forthcoming, I suggest that there is no reason why the evolutionist should not hold the law 'Whenever we have an instance of s, we have an instance of E'. Instances of s and E are not unique in any objectionable manner.

Of course, I would admit that it is most unlikely that an evolutionist would actually invoke so sweeping a law—however, the reason for this is because of the complexity of his material, not because there is any logical objection. He would want to spell out his *explanans* in detail. Nevertheless, if once he did this, by combining premises he could make such an all-embracing law. Furthermore, that the evolutionist would appeal to such a law seems to me to be far more plausible than that he would ever attempt the course of action suggested by Goudge. I am claiming that the evolutionist would set up his explanation in the following manner.

> *Explanans*   Instance of *s*.
>
> Whenever we have an instance of *s*,
> we have an instance of *E*.
> _____
>
> ∴ *Explanandum*   ∴ Instance of *E*.

Goudge's alternative is:

> *Explanans*   *s*.
>   _____
>
> ∴ *Explanandum*   ∴ *E*.

Now clearly both of us must be relying on some rule of inference to go from the *explanans* to the *explanandum*. The rule used in my formulation is the obvious one of *modus ponens*, a rule common to anyone who uses formal arguments. Goudge admits that some rule is necessary for his formulation and claims that the evolutionist is committed to the logical principle 'if *s*, then *E*'. In other words, he argues that the evolutionist will use the rule that if one gets *s*, one is allowed to claim *E*. (In clarification of his position Goudge cites Ryle's well-known paper ' "If", "So", and "Because" ' where it is claimed that 'When I learn *"if p, then q"*, I am learning that I am authorized to argue *"p, so q"*, *provided that I get my premiss "p"*.') But, even if Goudge were right in his claim that 'Whenever *s*, *E*' could not be a law, he still has the task of showing why on earth the evolutionist should be justified in using so esoteric a rule. This he does not do, and given the fact that *s* is (supposedly) unique, it is difficult to see how he could do it. (By his uniqueness argument he has ruled out any appeal to laws.)

In short, Goudge has argued that *s* is a set of conditions which is unique and unrepeatable, and yet he would also claim that within the evolutionist's logical apparatus waiting for just this occasion, there is the rule that if one has *s* then one can immediately claim *E*, a rule quite unknown to and unused by any logician or other scientist. The truth of such a claim seems so implausible that for this reason, if for no other, Goudge's position should be viewed with grave suspicion. For myself, assuming that Romer's explanation is typical of the kind of evolutionary explanation of concern to Goudge, I find Goudge's narrative model quite inadequate for an understanding of evolutionary explanation.

5.4   *Concluding thoughts on evolutionary explanations*

Obviously, a more important task than that of showing Gallie and Goudge to be wrong is that of giving the right model of evolutionary explanation (together with convincing reasons for why one should accept it). I have argued that I can see no good reasons for rejecting the covering-law model, and further, I have tried to show the difficulties which arise when one dispenses with the model's two major requirements—the aim for sufficient (or near sufficient) conditions, and the use of laws. Admittedly, it might be argued that I have not shown that a satisfactory explanation *must* contain a sufficient condition in the *explanans*; but I do think I have shown that one must go some way (in some sense) in this direction. I must confess that I cannot myself see how one could draw back from going a long way in this direction. The only way of avoiding this conclusion would seem to require an argument demonstrating that some necessary conditions are more important than others (together with the claim that important necessary conditions can suffice for adequate completed explanations). However, the difficulty is defining 'important' in this context without an appeal, implicit or explicit, to sufficiency. No one has done this and I shall shortly be giving three reasons showing why intuitively I doubt that it can be done. For this reason I support the covering-law model at this point, and it is also partly for this reason that I argue, as I did in Chapter 4, that evolutionary theory is a hypothetico-deductive sketch.

Together with my arguments about the relevance of sufficient conditions I think I have shown the implicit reference to laws (particularly those of population genetics) in all of the examples of evolutionary explanations that the covering-law model's critics have offered, and I have shown how, without laws, one must rely on esoteric rules of inference. However, even putting this point aside, to be honest I think that if some philosophers read a few technical evolutionary studies (rather than popularized accounts put out for the non-scientific public), they would find that many references to laws are explicit rather than implicit—they certainly are in the work of men like Simpson and Dobzhansky.

Finally, I have tried to point out that much of the difficulty with evolutionary explanations stems from ambiguities in the term 'unique'. When the critics of the covering-law model argue that evolutionary phenomena are unique, they do not realize that the senses in which these phenomena are unique are too weak to bear the conclusions they want to draw.

I personally find these arguments of this chapter, arguments which are certainly not very original to me, quite convincing. Moreover, to tie up loose ends left over from the last chapter I would suggest that we can now see even more fully how evolutionary theory presupposes and is unified by population genetics. The explanation of the evolution of the amphibians stands in exactly the same relationship to population genetics as does the explanation of Darwin's finches. However, I am sure that many readers will still feel less than overwhelmed by my logic. Possibly, they will feel for instance that, although the examples I have considered all involve an appeal to laws, logically it is still possible to have evolutionary explanations without laws. (Scriven, 1959, for example, denies laws a role in evolutionary explanations, although he does allow the place of a weak relative of laws.) More importantly and perhaps more generally, there may still be the nagging suspicion that, although any arbitrary specification of necessary conditions will not do, there comes a point when specification of additional necessary conditions is not needed (for a good explanation). Hence, before I conclude this chapter I would like to give three additional reasons why I find it intuitively more acceptable to suppose that the covering-law model, rather than any other model, is the correct guide to evolutionary explanations. My intuition's first source is the fact that evolutionists do actually offer covering-law explanations on some occasions. As we have seen, for relatively simple phenomena like instances of balanced polymorphism, explanations are provided which contain laws in the *explanans*, and the inferences from *explanans* to *explananda* are even made deductive. If evolutionists never gave covering-law explanations (or never explicitly appealed to laws in what I have called their 'sketches'), then I think the critics' case would be much stronger. Secondly, as we saw at the beginning of Chapter 4, central to an understanding of modern evolutionary theory is the premise that large-scale evolutionary changes involve nothing which is not to be found in small-scale changes. Admittedly, even if new principles were involved in large-scale changes, this would not necessarily mean that a new kind of explanation is required —but, on the other hand, since we are told that exactly the same forces are working in all cases, large or small, I myself just cannot see why the kind of understanding required in all cases should ever differ (and, I have just pointed out, the understanding offered in small-scale cases does seem to be covering-law understanding). Thirdly and finally, my intuition about the nature of evolutionary explanation comes from the knowledge that covering-law modellists can supply very good reasons why, in fact, evolutionists do not always give covering-law explanations. So much evidence required for the completion of covering-law explanations is missing, and even if it were available, it would just overwhelm evolutionists by its bulk. If all the needed evidence (and time) were available and evolutionists just turned their backs on it, then my sympathies with the critics of the covering-law model would be much stronger. As it is, my sympathies lie in the other direction, for, apart from anything else, as we saw in Chapter 4, when new evidence is discovered (e.g. the facts of heredity), this seems to have the direct function of increasing actual exemplifications of covering-law-type arguments in evolutionary studies.

Two final points need to be made. First, I have just given an argument based on the inference from parts to wholes—from small-scale evolutionary changes to large-scale changes. It might be thought that, attractive though such arguments are, it is just not true that large-scale evolutionary changes involve nothing occurring in small-scale changes. For instance, one can meaningfully talk of a 'trend' lasting millions of years (say, from small size to large size). However, if one got change in a couple of generations, this would be no 'trend' but (at best) a 'jump'. Hence, some things necessarily occur on the big scale but not on the small scale.

I do not think this criticism is very worrying. After all, in physics one gets an analogous situation; but no change is needed in the form of explanation. If a planet goes two feet it is hardly describing an 'orbit', although the form of the explanation for any distance of travel is the same. The truth is that things like trends and other 'wholes' occurring only in large-scale changes can be analysed completely in terms of 'parts' occurring in small-scale changes. A trend is no more than the sum of a number of small changes spread over a length of time. But, if the parts can be explained by a covering-law explanation, then logically it follows that the conjunction of the parts *can* also be so explained—the question is whether they *need* to be so explained.[2]

My second point is that, without wishing to retract in any way, I would mention that in this chapter I have argued the stronger of two options open to me. I might only have argued a claim that follows from my claim—rather than arguing that evolutionary explanations are in some sense covering-law explanations, I might have claimed merely that evolutionary explanations do not differ essentially from explanations in the physical sciences. Many philosophers believe strongly that the covering-law model is not always appropriate even in the physical sciences. To them I offer the weaker option (although I think the stronger is there to be taken), and I would suggest that whatever else we may have seen, we have seen nothing to justify a division between evolutionary explanations and physical or chemical explanations.

To write more at this stage on explanation would be pointless. Let me therefore turn to my final chapter directly concerned with evolutionary theory, wherein I shall raise a problem closely linked to some points which have already been raised.

[2] Note that I am not here saying that every biological whole is no more than the sum of its parts—in particular, I am not saying that when the parts are molecules, the whole is the sum of the parts. What I am saying is that for things like trends, where the parts are biological, the whole is no more than the sum of the parts.

6

THE THEORY OF EVOLUTION

III: EVIDENCE

Almost without exception living evolutionists accept the synthetic theory of evolution, the theory of evolution discussed in the preceding chapters. This does not mean that they accept without a murmur all of the claims of the theory—indeed, as I have tried to point out, in one way no part of the theory is immune from attempted change. Genetics, for example, feels the effects of molecular biological studies (as will be discussed in Chapter 10) and population geneticists themselves are trying to eliminate some of the crudities of their earlier efforts—for instance, they work assiduously to get away from 'bean bag' genetical thinking which treats genes (falsely) as independent units in the cell, each gene affecting the phenotype in isolation from all others. (But see Haldane, 1964.) Moreover, some parts of the synthetic theory are still highly controversial—recent disputes have centred on such things as the possibility of genetic drift, the importance of the founder principle, the possibility of reproductive isolation without (at some point) having geographical isolation and the question of whether selection can ever be for the group rather than for the individual. (Williams, 1966, discusses many of these disputes. At a more particular level, Bowman, 1961, has challenged several of Lack's claims about Darwin's finches.) Nevertheless, despite this debate (which is of a kind common I am sure to every living science), I think one would be hard put to find an evolutionist who would claim that the essential outlines of the true mechanism of evolutionary change are still unknown. Evolution is, to put it simply, the result of natural selection working on random mutations.

In the face of this unanimity, it is therefore a little surprising that many who comment philosophically on the theory think that its basic truth is still open to grave question. There does not seem to be any one reason alone behind this questioning of the theory; but I think that a primary barrier in the way of the theory's acceptance is a fairly commonly held view about the nature of theory-confirmation. It is felt that if we are to confirm a theory T, that is, to find supporting evidence in T's favour, what we must do is to make some predictions on the basis of T. If the predictions turn out to be true, then this is evidence in favour of T; if the predictions turn out false, then this is evidence against T. However, conjoined with this view is the fact that (apparently) evolutionary theory has a very low predictive power. Whatever one's view of the right model of evolutionary explanation, no one could pretend that we can predict the future evolution of, for example, elephants. And even if we could, no one would be around to check out the predictions. Hence, neither in principle nor in practice can one falsify the theory, and conversely, genuine evidence cannot be found in favour of the theory.

Thus, for example, as we have seen, Manser argues that all the theory offers us is a picture of the process of evolution, and he adds that predictions are impossible, and that hence, evidence for the theory is nonexistent. He thinks that the reason why it is accepted is, apart from the fact that it is analytic, that it 'has never had to deal with serious scientific opposition' (Manser, 1965, 18). Even when opposition has been raised, it (the opposition) has been down-played, 'possibly not on very adequate evidence' (Manser, 1955, 18). Goudge, although much more sympathetic to the synthetic theory, thinks that at best evolutionary theory gives one negative predictions telling one what will *not* turn up (Goudge, 1961, 78). Far from being concerned about the lack of opposition to the synthetic theory, he seems to think that one of its rivals, the so-called 'saltation' theory, could well be true. 'A philosopher surveying the pros and cons of this controversy finds himself in no position to espouse one side rather than the other. The only reasonable conclusion seems to be that considerations so far advanced do not permit a settlement' (Goudge, 1961, 51). Himmelfarb argues that a Darwinian theory (i.e. an evolutionary theory based on natural selection) holds only because Darwin invented a new logic, a 'logic of possibility'. 'Un-like conventional logic, where the compound of possibilities results not in a greater possibility, or probabilty, but in a lesser one, the logic of the *Origin* was one in which possibilities were assumed to add up to probability'(Himmelfarb, 1962, 334). Barker (1969), by judiciously picking out what he takes to be common ground between Darwin's theory and the modern theory, offers a somewhat incred-ible transcendental deduction of the fact, not only that the theory

D

(whose?) is unfalsifiable, but that it can have no rivals. And Carlo writes that: 'The evidence in favour of evolution theory has never been completely satisfying. Strictly speaking, evolution has yet to be scientifically proven' (Carlo, 1967, 120).

My aim in this chapter is to show that these critics are quite wrong. I shall begin by discussing the evidence in favour of the central body of the theory—in particular, I shall look at the evidence from the study of wild populations of organisms and then at the evidence from captive populations. Next, I shall consider three rival theories of evolution, and shall evaluate these and the synthetic theory through the evolutionary areas of morphology, systematics, and paleontology. Finally, I shall consider briefly why some of the critics go astray. By the time I have finished I hope I shall have shown that evolutionists' confidence in their theory is fully justified, although I shall try not to minimize areas of debate. Indeed, part of my plan is to show the magnitudes of the problems encountered in evolution experiments—problems which often make it extremely difficult to come to a definitive decision about particular matters of evolutionary controversy existing within the confines of the synthetic theory.[1]

6.1 Heritable change in wild populations

We know now how genetics—mainly population genetics, but more generally Mendelian genetics—is central to the modern theory of evolution. Evidence in favour of Mendelian genetics, therefore, is evidence in favour of evolutionary theory. The evidence in favour of the central parts of Mendelian genetics (i.e. the parts dealing with the individual) is of two kinds, that which I have called direct, namely the cytological evidence, and the indirect, the evidence that we have of the transmission of phenotypic characters. Now, I admit (as before) that some of the laws of genetics, Mendel's laws in particular, have exceptions; but (again as before) I would point out that these exceptions are no more than one finds in the physical sciences and that, by and large, it would be impossible to think of stronger or more widespread evidence in favour of a theory than there is in favour of those parts of Mendelian genetics dealing with the individual. Therefore, since the central area of Mendelian genetics seems relatively unproblematic (from the viewpoint of our present discussion), let us turn to the area where more interesting problems arise, namely population genetics.

[1] Excellent discussions of the evidence for evolutionary theory can be found in Ford, 1964, and Dobzhansky, 1970, although these books taken together illustrate perfectly the basic points I want to make in this chapter. Neither author has any doubts about the fundamental claims of the synthetic theory— natural selection working on random mutations—but they differ significantly over the tenability of specific details within the theory.

We have already seen some evidence in favour of what is held about population genetics, for something like Race and Sanger's study of blood groups obviously provides support for the Hardy–Weinberg law (although I think that most biologists would feel that the major support for the H–W law comes indirectly from Mendel's laws). However, there is another type of evidence about population genetics which has not yet been discussed, namely the evidence of *heritable* change occurring in groups over periods of time, where the situation is, as yet, often too complex for a detailed formal analysis. In this section, let us ask what evidence the evolutionist might have of such heritable change occurring in populations of wild organisms. In the next section, let us ask what evidence controlled experiments on populations in, for example, laboratories and zoos can throw on the basic problems of heritable change in groups.

Now, as soon as we start to consider the question of the study of heritable change in wild groups, because evolutionary theory claims that significant heritable change tends to be slow and gradual, a number of factors become relevant. For instance, unless one has organisms reproducing fairly rapidly, one is not likely to notice much in the way of heritable change—hence, organisms with only two or three generations per century can probably never be profitably studied (unless, for some reason, very detailed records are kept, as in the case of man). Also, it helps if the organisms are under severe selective pressures—the more devastating the lack of the right adaptations, the more rapid the spread of the fitter genes. Nevertheless, despite factors like these which tend to reduce the likelihood of re-cording evolutionary change, it does seem that enough favourable circumstances have combined on occasion to enable evolutionists to find quite unmistakable evidence of heritable change occurring in wild populations. And moreover, this is heritable change which is, as the synthetic theory predicts, primarily a function of natural selection. Let us look in some detail at the most famous case, that of industrial melanism in moths. (Full details can be found in Ford, 1964.)

A hundred years ago, the Peppered Moth, *Biston betularia*, was almost entirely speckled. However, as the years passed, in industrial areas a melanic (i.e. black) form of the moth became increasingly common, until now in some places it is the usual form. Since it is known that this melanic form is due basically to a single dominant gene, here we clearly have a heritable change occurring in a species. Moreover, the cause of this change, long suspected, is now definitely known. One of the biggest dangers the moths face is that of being eaten by predators, namely birds. When trees are clean, the speckled moth is better camouflaged against lichens or tree-bark than a melanic moth. On the other hand, when trees are dirty from soot, the melanic moth is at an adaptive advantage. Hence, since through the industrial revolution, trees (in cities) got dirtier, the balance of fitness changed from the speckled moths to the melanic moths, and there was consequently an evolution from the former to the latter.

This phenomenon which has occurred in the moths, even if one were to concede that its overall significance is limited, seems to me to be quite clear-cut evidence in favour of the major tenets of population genetics, and hence evidence in favour of the synthetic theory of evolution. However, at least one philosopher, Manser, thinks otherwise for he argues that the account just given of the evolution to the melanic form 'is only a description in slightly theory-laden terms which gives the illusion of an explanation in the full scientific sense. If no mutant forms had occurred and the species had become extinct as a result of the change of circumstances, it would not have been adaptable. We cannot use the account to predict what will happen when a new feature occurs in the environment of a different species, or even if there is another change in the environment of the original moths. All we can do is, after there has been time for the state of affairs to become stable again, say whether the species in question was or was not adaptable' (Manser, 1965, 25-6). For these reason, Manser seems to think that the example has been moulded to fit the synthetic theory of evolution. It provides no true test of the theory; but conversely, can give the theory no empirical support.

There seem to be two questions raised by Manser's criticism. In the first place, can events which have already passed ever provide evidence for a theory (or can events which were only discovered when they were past support a theory)? Must a theory, therefore, rely for support purely on predictions about the future? If this is the case, then the moth example is certainly no evidence for evolutionary theory, nor is any like example. Secondly, is there something suspect in itself about *this* example of the supposed evolution of the moths? Is there some reason why this example cannot provide genuine empirical evidence in favour of the theory of evolution?

The answers to neither of these questions support Manser's position. In other words, we do have genuine evidence here in favour of the theory. Consider the first question. Obviously past events can be used to confirm a theory. For example, if, with the aid of Newtonian theory, we infer backwards to an eclipse whose existence is then confirmed by historical records of reputable eye-witness reports, this is just as good evidence for the theory as an inference forward to an eclipse whose existence we later confirm through eyewitness reports. Confirmation (or refutation) of theories does not rest on the exact time at which events occur—rather it rests on the relationship between event and theory. Specifically, does a description of the former follow in some sense from the latter? (This is not to deny that our faith in the latter is probably increased if the former is unexpected.) Hence, for this reason, there is nothing suspect about using the change in the moths as evidence for the theory of evolution. As far as the second question is concerned, Manser's doubts about the worth of the example under discussion seem to stem from his mistaken notion that natural selection is fundamentally and inescapably tautological. He seems to think that because the black moths increased in number, analytically they are defined as the fitter without further argument. However, despite what Manser thinks, whilst it was indeed the case that because the black moths seemed the more successful, evolutionists suspected they were the fitter, the black moths were not circularly defined as the fitter and the matter left at that. Evolutionists went out to find the causes behind the black moths' apparent fitness, something quite unnecessary if all that is at stake were a definition. Kettlewell (1955) ran several tests showing beyond reasonable doubt, both that a chief source of danger to the moths is being eaten by birds, and that when trees are soot-covered, the black moths' camouflage gives them a selective advantage. In other words, he proved both the existence of a differential survival (from which one can draw inferences about a differential reproduction) and that the sorts of things which increase fitness in other circumstances are the things increasing fitness in this circumstance (in particular, camouflaging characteristics). Hence, notwithstanding Manser's claims, it would seem that we have here an example which supports both the facts of evolution in general and the synthetic theory in particular, because it was shown that there is the kind of uniformity which serves as the basis of inductive generalizations, which, as I pointed out earlier, lie behind the evolutionists' empirical claims about selection and adaptation.

This case is not unique. Numerous similar cases are known. For example, rabbits have developed a genetic immunity to myxomatosis, viruses develop resistance to drugs (indeed, some viruses can now exist only *in* the presence of drugs once introduced to eliminate them), and it seems fairly clear that some evolution has even occurred in man recently (Dobzhansky, 1962). In particular, because of the deadly effects of tuberculosis (intensified by the move to city living), in the past 200 years man seems to have evolved into a form more resistant to TB (by virtue of the fact that those most liable to TB had less offspring). However, in all of these cases, including that of the moths, too much should not be claimed. In particular, the amount of evolutionary change involved is very small—indeed, in the case of the moths this change is not much more than the minimum of one gene. Hence, were someone to put forward a theory (which, as we shall see shortly, someone has in fact put forward) allowing for the possibility of small change due to selection, but denying that large-scale changes have the same cause, evidence of the type we have just been considering would be unable to decide between the synthetic theory and it. Possibly this kind of situation is sufficiently common to be worth noting. In many areas, the evidence in favour of one theory (particularly the synthetic theory) and against each and every one of the other theories is not always overwhelming. It is only when one considers the overall evidence, and most particularly the central claims of genetics, that one recognizes that the synthetic theory alone receives positive support from all evolutionary areas of study. As we shall see, the other theories all get insurmountable checks from at least one area (for a start, they all make some untenable claims about genetics), and they frequently say nothing about other important areas.

6.2   *Heritable change in captive populations*

In order to discuss the evidence which comes from the study of captive populations, it is necessary first to draw a distinction between *artificial selection* and *natural selection*. Both kinds of selection involve differential reproduction; but whereas natural selection is entirely a function of the characters of organisms and of the reaction of organisms to each other and the environment, artificial selection occurs only when one has a rational agent consciously trying to influence the chances of one organism reproducing rather than another. It is the kind of selection which occurs when the breeder tries to improve his stock or the fancier tries to improve the particular species of organism which is his hobby. Obviously, one can have natural selection without artificial selection; but since human plans often fail and prize animals turn out to be infertile, artificial selection is probably usually accompanied by some natural selection.

Bearing in mind this distinction between the two kinds of selection, there are, I think, three questions which are pertinent to the problem of the evolutionary evidence to be gleaned from captive populations. First, can artificial selection, done purely for practical or aesthetic reasons, nevertheless support the claims of population genetics and, more generally, of evolutionary theory? Secondly, can one devise experiments using artificial selection which will support the synthetic theory and natural selection? Thirdly, can one achieve natural selection in artificial surroundings, and if so, what support can this give to the synthetic theory? Let us take these questions in turn.

As far as the question of the worth of artificial selection, practised purely for profit or pleasure, is concerned, it cannot be denied that it has certainly been thought that such selection has relevance for at least one evolutionary theory based on a form of natural selection. Charles Darwin devoted the whole of the first chapter of his *Origin of Species* to a discussion of such instances of artificial selection, arguing that they strongly supported his theory. For example, about one instance of the breeder's art (pigeons) he wrote: 'Altogether at least a score of pigeons might be chosen, which if shown to an ornithologist, and he were told that they were wild birds, would certainly, I think, be ranked by him as well-defined species. Moreover, I do not believe that any ornithologist would place the English carrier, the short-faced tumbler, the runt, the barb, pouter, and fantail in the same genus . . .' (Darwin, 1959 ed., 97). Yet, simultaneously, Darwin was able to show that in fact all of these different breeds of domestic pigeon are descended from one and only one species of wild pigeon. Because of this, he argued that since artificial selection can have so drastic an effect on the characters of a population of organisms, it is reasonable to suppose that natural selection can have a similar effect. He asked: 'Can the principle of selection, which we have seen is so potent in the hands of man, apply in nature? I think we shall see that it can act most effectually' (Darwin, 1959 ed., 163).

It seems clear that Darwin was right in citing the diversity of the pigeons as evidence for his theory and we would be right in citing the pigeons as evidence for our theory based on natural selection—the question is just how far the value of such evidence extends? To answer this, let us consider the similarity between artificial selection and natural selection. In both cases, we have only a fraction of the potential reproducers actually reproducing and having fully viable offspring. But what these cases of artificial selection show is that, even though the differences between the successful and unsuccessful organisms may be very slight, the cumulative effect of this differential reproduction can be very great. Hence, given differential reproduction, there is no need to hypothesize the existence of large variations to account for large overall changes. However, since natural selection is also a type of differential reproduction, it therefore seems legitimate to argue for this conclusion for natural selection on the basis of artificial selection. In other words, artificial selection shows that all kinds of selection (including natural selection) could lead to large changes, even though only small variations were involved.

Nevertheless, this said, the analogy between these instances of artificial selection and natural selection breaks down, In particular, one cannot argue (from these instances) that natural selection does ever occur, or if it occurs that the kinds of variations to be selected will be the same as those selected by breeders. Because the breeder of Afghan hounds selects for long coats, it does not necessarily follow that there will be selection for long coats in wild populations of dogs. Consequently, the evidence from these cases of artificial selection is very limited. They show that selection can have big effects; but they do not show either that natural selection exists or that if it does exist, in what direction such selection points.

Next, we have the question of whether or not a selection performed by man can be used to support evolutionary theory when the deliberate intention is to find evidence in favour of evolutionary theory. (This would be evidence from artificial selection over and above that just discussed when there is no intention of supporting evolutionary theory.) It seems fairly clear that such evidence from artificial selection can be found, if one can give good reasons for suggesting that one is filling some of the variables left empty in the cases mentioned above (i.e. cases where artificial selection is practised only for gain or personal delight). In particular, evidence would seem to be forthcoming if one can suggest that one's artificial selection bears analogies (in addition to differential reproduction) to possible cases of natural selection. If this is so, then one might be able to learn the directions which natural selection could take, the rates of its effects, and some of the overall consequences.

A good example of where evolutionists have used artificial selection to try to simulate natural selection in this way has occurred in experiments attempting to solve a problem already mentioned which has long provoked and divided evolutionists (a problem, that is, which has divided holders of the synthetic theory—not one which poses a challenge to the foundations of the theory itself). Some evolutionists, particularly Mayr (1963), argue strongly that speciation in higher animals (the dividing of such animals into groups which are reproductively isolated from each other) cannot occur without geographical isolation. They argue that unless groups are geographically separated for some time, the groups cannot build up the genetic foundations of reproductive isolation. Other evolutionists, for example Ford (1964), whilst agreeing that geographical isolation usually precedes reproductive isolation, feel that possibly one might sometimes get reproductive isolation without geographical isolation. Thoday and his associates (e.g. Thoday and Gibson, 1962) have carried out a series of selection experiments (i.e. experiments wherein artificial selection is trying to copy natural selection) in order to throw light on this problem. Starting with small populations of *Drosophila*, Thoday removed those members with certain characters. (The characters involved were chaetae. Thoday removed those with intermediate numbers, and left those with high and low numbers.)

After a very short time (10 or 12 generations) Thoday was left with two reproductively isolated populations, even though there had been no geographical isolation and it was deliberately ensured that there would be some interbreeding between high and low chaetae-carrying flies. Hence, it would seem that artificial selection can achieve reproductive isolation without geographical isolation, and assuming that one gets roughly analogous conditions in nature, natural selection can do the same.

Nevertheless, it should be noted that although some evolutionists do think that such analogous conditions will sometimes obtain, the analogy, as in most cases where one is trying to use artificial selection as a substitute for natural selection, is at points somewhat tenuous. For a start, although very severe selective pressures do occur in nature, the experiment does not really show what kind they might be—it is not necessarily the case that *Drosophila* chaetae might be the particular characters involved, although of course they could be. And many would argue that, other than under very unusual circumstances, such severe selection pressures would be unlikely to hold for long enough for anything but very rapidly reproducing organisms like insects (although there is strong evidence that such forces commonly hold amongst insects). Secondly, the captive populations were separated from all other species, and more importantly, from other groups of the same species. Mayr (1963) estimates that in any population one might get up to a 40 per cent turn-over due to emigration and immigration, although recently, Ehrlich and Raven (1969) have challenged this assumption. Thirdly and perhaps most importantly, the experimental populations were very small—in the region of about 40 members. None of these points seem to show absolutely that the analogy between Thoday's experiments and possible wild populations could never hold; yet they do suggest that if ever it does occur, speciation without geographical isolation in higher animals will be rare, and it will possibly be restricted to organisms of particular kinds (like insects).

Probably most selection experiments are like Thoday's in that one must make qualifications about the conclusions one can draw. Such experiments obviously throw some light on the problems of evolution; but, taken on their own, the experiments would rarely seem to end debate entirely. (Thoday's experiments have certainly not been the final word in the debate about speciation.) However, even if one grants this, at this point some critics might complain that the whole discussion so far has been an evasion of the main point. The question which should be asked, these critics would say, is the third one posed above, namely, what evidence can one give of the existence of *natural* selection by using captive populations? If we do not first establish the existence of natural selection, then do not all of our analogies from artificial selection become futile? Let us therefore turn to this question, noting however, that even if one were able to achieve no natural selection in captive populations, we still have evidence for it in wild populations.

There certainly seems no reason, in principle, why natural selection should not occur and be studied in captive populations. Given organisms' geometric capacity for population increase, then if one restricts this capacity by regulating food or space or something, then many are going to die without reproducing. Moreover, success and failure will be a function of the characters possessed by the organisms —it will not be a function of the experimenter's decision to let some succeed and others fail, as it would be were artificial selection occurring. A good example of an experiment involving this kind of natural selection in a captive population is given by Dobzhansky (1951). Populations of *Drosophila melanogaster* were kept in cages —the populations consisting of mixtures of the wild type (i.e. flies with normal-length wings) and of a mutant, 'vestigal' (i.e. flightless flies with stubby wings). When no special adjustments were made to the environment, the wild types increased rapidly, pointing to the fact, not only that natural selection was occurring, but that the wild-type was fitter. However, one might suspect that in some circumstances the vestigal would prove to be fitter, and this turned out to be the case. On small oceanic islands a lack of wings could be adaptively advantageous, for winged insects would tend to be blown out to sea and destroyed. Experiment confirmed this supposition, for when the captive populations were kept in places where wind could blow members away, it was the vestigal which proved fitter and increased in number.

No more need be said about the value of experiments like these. They supply evidence not only of the existence of natural selection, but also of the effects of the sorts of selection we might expect to find in nature. However, here as before, we must temper our enthusiasm with some qualifications. For a start, once again in one's studies one is restricted to rapidly breeding organisms. Secondly, we must recognize that some organisms just get quite disoriented by captivity—some refuse to breed in captivity whereas they would in nature, and conversely, some normally reproductively isolated organisms hybridize madly in captivity, producing fully fit offspring. Thirdly, one must face up to the fact that, try as one might, captivity is going to create some new conditions and possibly new selective forces. Thus one cannot really hope to study the effects of very slight forces, because as selective forces get milder, the possibility that there are other, unplanned forces, also having effects, gets greater.

Since, it is thought that evolution is often the product of very weak forces, this is obviously a severe limitation on natural selection experiments.

In conclusion, it would seem that in all the cases we have studied the answer is similar. One can get evidence—substantial evidence—from the study of captive populations; but, there are limits to the extent of this evidence. Nevertheless, even though this is so, it is important to end this section by pointing out one limitation which is not relevant here. It might be objected that most of these experiments on captive populations are of very dubious value, because most of them do not pretend to tell us the actual course of specific evolutionary paths. They aim to tell us only how evolution could occur. However, whatever may be the limits of experiments on captive populations, this objection does not point in the direction of any of them, for the criticism is based on a confusion revealed earlier, namely that between a *theory* of evolution and a *phylogeny*, that is a description of a group's evolutionary path. The experiments are designed to tell one about the theory—this they do. They are not aimed primarily at the actual reconstruction of the history of life, and for this reason should not be blamed when they do not tell us about it. (Of course, this is not to deny that one might apply the results of one's experiments to actual situations.)

6.3   *Alternative evolutionary theories*

Manser's claim that the theory of evolution through natural selection has never had serious scientific opposition could not be more wrong. At one time, during the early development of Mendelian genetics, many of the leading biologists thought it was an outmoded, dying theory, and until recently, not a few first-class biologists held alternative theories to the synthetic theory. In this section, I shall look briefly at three—(i) so-called 'Lamarckism' (Lamarck, 1809; Cannon, 1955; 1958; Mayr, 1972), (ii) the 'saltation' theory (Goldschmidt, 1940; 1952; Schindewolf, 1950) and (iii) the theory of evolution by 'orthogenesis' (Jepson, 1949).

By *Lamarckism* is commonly understood the theory that evolutionary change is a function of the direct effects of the interaction of the individual organism and its environment—heritable characters appearing and disappearing through the effects of use and disuse. However, I think that 'true' Lamarckism (i.e. the version held by Lamarck), although including use and disuse, is rather broader and covers all kinds of evolutionary change which are supposedly the result of heritable characters appearing or disappearing in response to 'needs' brought through environmental pressures. (This excludes random mutation which is hardly a response to pressure.)

Indeed, 'true' Lamarckism probably puts the greatest emphasis on some kind of general tendency of all organisms to evolve towards a kind of perfection, although Mayr (1972) suggests that as his thought developed, Lamarck came less and less to push this to the fore of his evolutionary discussions.

Prima facie, Lamarckism (*qua* response to need) seems quite plausible, for there are many characters which are apparently best explained in a Lamarckian fashion. For example, friction on (human) skin causes calluses—hence, the horniness of the blacksmith's hands. However, human babies are born with thick skin on the soles of their feet. What more natural supposition could there be than that once babies were born with thin-skinned soles, and that the thick skin only developed after they began walking on their bare feet. Then, because of this, future generations started to inherit what earlier generations had to acquire. Other characters like this are the black skin of humans (acquired through generations of sun-tanning), the calluses on the backsides of ostriches (acquired through generations of sitting), and, of course, the long necks of giraffes (acquired through generations of stretching). Apparently one can even find experimental evidence of Lamarckism. The geneticist, Waddington, subjected *Drosophila* larvae to heat-shocks and found that some of the flies then developed with wing deformities. He selected these flies with deformities, and after a few generations he got flies with wing deformities even *without* heat shocks. Here, seemingly, is a clear-cut case of a Lamarckian-type heritable change (Waddington, 1957).

Of course, one major problem with Lamarckism, a problem even its supporters recognize, is that of filling in the genetical background which is supposed to be responsible for the transmission of characters which appear (or disappear) from one generation to the next through use and disuse or just from a general need. One who tried to remedy this deficiency was Darwin (1868), who, believing in the heritable effects of use and disuse, offered what he called a theory of 'pangenesis'. Darwin supposed that all of the cells of the body throw off minute gemmules, that these are carried around the body, and that they are then transmitted via the sex-cells to the next generation. He suggested that an alteration in the physical characters of an organism through use and disuse causes an alteration in the gemmules, which are then passed on in an altered form to the next generation, who thus inherit the changed physical characters. Hence, we get a Lamarckian effect.

Nevertheless, attractive though Lamarckism seems at first sight, there are several insurmountable reasons why both Lamarckism and Darwin's pangenesis are considered untenable today. The most obvious and most basic is that the cytological evidence points entirely in another direction. Cells do not give off loose units of heredity just like that, there are no pangenes carried along in the blood or in any other bodily fluid, there is absolutely no (cytological) evidence that the environment has a direct effect on the units of inheritance either through use and disuse or from a general need in the way supposed by Lamarckism—the evidence all points the other way, namely that when the environment affects the genes it does so through such things as radiation, and the phenotypic effects are quite unrelated to organic needs or use and disuse—and, finally, it is just not the case that genes or gemmules come from all over the body to form the sex-cells ('gametes'). The formation of these cells in many organisms is quite separate from the other ('somatic') cells. In woman, for example, those cells which form the gametes are already in the ovaries before birth. Hence, in no way could use or disuse or any kind of need brought on by the environment have a heritable effect.

These reasons, contrary to Manser's unbelievable claim that Lamarckism has not been disproved on 'very adequate evidence', make belief today in Lamarckism (and its corollary, pangenesis) on a par with belief in a flat earth. However, for some reason, Lamarckism continues to exert a fascination on laymen, and at periodic intervals its 'confirmation' is triumphantly heralded, usually on the basis of some esoteric experimental results which prove something quite undreamed by Lamarck, Darwin, or any other serious scientific upholder of the doctrine. (One favourite example is the effects of transferring DNA from one organism to another.) But rather than trying to fathom the strange psychology of neo-Lamarckians, let us instead conclude this brief discussion of Lamarckism by turning back to Waddington's experiment.

First, it should be noted that Waddington's effect is quasi-Lamarckian at best. There is no evidence that the wing deformity was any kind of response in answer to a *need* set up by the heat-shock. Secondly, even for this borderline Lamarckian effect, Waddington did not have to step outside of the bounds of orthodox Mendelian genetics to find an explanation. Waddington's (Mendelian) explanation started from the fact that organic characters are always a function both of the environment and the genes. In particular, the deformity of the flies which needed heat-shock has to have a genetic basis as much as it has to have an environmental basis (i.e. the heat-shock). Waddington supposed that there are a number of polygenes, each capable of creating deformity when conjoined with heat-shock, but unable to create deformity when alone and without heat-shock. He selected experimentally for those genes, and so his final specimens carried several such genes which Waddington hypothesized could together cause the deformities even without heat-shock. To test his hypothesis, Waddington argued that if it were true, then by natural accidents of sampling, occasionally even without selection a fly would accumulate enough polygenes to exhibit wing deformity. This in fact occurs, and hence it would seem that in order to explain the effects of the experiment, there is no need to suppose any kind of heritable mechanism other than that supplied by Mendelian genetics. A Lamarckian hypothesis is unneeded.[2]

An evolutionary theory which was popular both with geneticists and paleontologists in the early part of this century was the *saltation* theory of evolution. It is this theory which Goudge, unlike active biologists, still thinks is a viable theory of evolution. The basic premise behind the theory is that although intra-specific change can involve just selection working on mutations having small heritable effects, large-scale evolutionary changes call for instant, large-scale, heritable changes. These evolutionary jumps are supposedly brought about by 'macromutations'. Hence, it is claimed, we get an instant, one-generation jump from, say, reptiles to mammals, and then, in turn, we get a jump from the primates' immediate precursors to the primates, and then from man's immediate ancestors to man. There are no transitional stages—one is either a reptile or a mammal, one cannot be something in between. Since these changes are so rapid, there is hardly place for the effects of natural selection, and so it is argued by some that one cannot expect to see moves in the direction of greater adaptation in the effects of macromutation. In other words, the basic 'plans' of organisms—that which makes a reptile a reptile rather than a mammal, and a man a man rather than a chimpanzee— are not adaptive. 'Adaptive' is a predicate which can hold only of differences between organisms of the same species.

I shall have more to say about this in the following sections when we come to consider the various subsidiary areas of evolutionary biology. At this point, let me note that, as in the case of Lamarckism, the theory fares badly in the light of the evidence of genetics and cytology. The theory owed its initial plausibility mainly to the fact that, not surprisingly, early geneticists concentrated primarily on fairly large, unequivocal instances of heritable change. It was not

---

[2] In fairness, it should be added that there is a rather paradoxical conclusion to this story. Waddington himself does actually think that his experiment points to a way to supplement the synthetic theory by a mechanism which he calls 'genetic assimilation'. Waddington thinks his experiment supports the hypothesis that very rapid evolutionary change can be caused by characteristics due to polygenes first being manifested through environmental stress and then being selected until the stress is not needed. However, Williams (1966) points out that there is no reason to suppose that stress will ever cause characteristics which will be favoured by natural selection—the wing deformities in question are not such an example.

until genetics had matured a little that it was realized just how widespread is the existence of mutations causing very small changes. However, although geneticists know of some mutations which cause fairly drastic changes, they have entirely failed to discover the kind of macromutations required by the saltation theory—the kind of mutation which would take a group of organisms from one order to another. Moreover, the large-effect mutations which are known are usually just those mutations which are the most crippling to their carrier from an overall reproductive viewpoint, although, contrary to the claims of the saltation theory, they do not make for reproductive isolation between parent and child (unless they end reproduction entirely). Achrondroplastic dwarfism is caused by a one-gene mutation and it certainly has fantastic effects on the phenotype. However, the human achrondroplastic dwarf is not even reproductively isolated from other humans, although his (or her) reproductive potential appears to be much reduced.

Of course, one might argue that the failure to find the right kind of macromutations does not necessarily prove their non-existence; but, like unicorns, there is a difference between saying that *logically* they might exist and that it is *reasonable* to suppose that they exist. Moreover, as Dobzhansky (1951) has pointed out, in order to hold the saltation theory one must suppose not only the existence of macromutations, but also that such mutations occur twice in the same individual (and, presumably, in at least one contemporaneous organism of the opposite sex). Otherwise, one will have organisms with just one such mutation, and the whole point is that the mutations are supposed to carry one from one reproductively isolated group to another—that is, from and to groups where hybrids (i.e. organisms with just one half set of genes from each group) are reproductively handicapped, if not outrightly non-existent (like a hybrid reptile-mammal). An organism with one macromutation would be just such a hybrid. Thus, to hold the saltation theory, one must believe, not just in the existence of macromutations, but that they can occur twice in the same organism (and, if they are to be transmitted, in a mate also). The need to suppose rates of mutation like this, given that none have ever been spotted, makes a belief in their existence ludicrous.

Nevertheless, as in the case of Lamarckism, there are some phenomena which prima facie support the saltation theory. In some groups of organisms, virtually all of which are plants, one does find that speciation occurs in the space of one generation (Stebbins, 1950). That is, as the saltation theory supposes, the parents and offspring are reproductively isolated. However, whilst this phenomenon, known as 'polyploidy', does show that the synthetic theory must be extended to cover such cases of speciation, it hardly provides support for the saltation theory as described above. Polyploidy is a function, not of macromutations, but of whole sets of chromosomes being passed on, rather than as is normal, haploid sets. Thus, instead of the offspring getting $2n$ chromosomes ($= n + n$), they might get $3n$ ($= n + 2n$) or $4n$ ($= 2n + 2n$). Or they might get whole sets from organisms of different species. The most famous example of the latter kind of organisms is the hybrid between cabbage and radish, which can reproduce but which is (reproductively) isolated from both kinds of parent. But, for most groups of organisms, polyploidy is rare or non-existent, and so, even if it did support the saltation theory (which it does not) it would still supply no answer to the missing macromutations supposedly involved in most evolutionary changes.

This brings me now to the third rival to the synthetic theory, namely evolution by *orthogenesis*. I shall say very little about this theory at this point, mainly because it says nothing very much about genetics or the causes of heritable change generally. It was argued that the fossil record gave evidence of trends (from smaller to larger and so on), and that towards the end of these trends the characters involved, whatever their earlier adaptive value may have been, became positively injurious to their possessors. For example, it was claimed that something like the massive fangs of the sabre-toothed tiger, or the monstrous horns of the Irish elk (see Figure 6.1), required an explanation which could not be supplied by any theory based on the Darwinian-inspired notion of natural selection. It was therefore concluded that one must recognize the existence of some kind of orthogenetic 'force' or 'momentum' carrying characters beyond the point of maximum adaptive value.

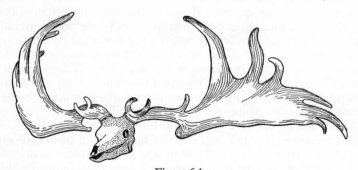

Figure 6.1
The horns of the Irish elk, *Cervus megaceras*. (From Carter, *Animal Evolution*, Sidgwick and Jackson Ltd.)

I shall deal shortly with the question of whether or not evolutionary theory implies that all organic characters (in successful) organisms must be adaptive, and also with the supposed trends found in the fossil record. At this point, let it be noted merely that the major reason why this theory of evolution by orthogenesis is ignored today is because it, like the other two theories discussed above, fails to answer the questions of genetics and cytology. Indeed, it does not even really try to answer them in any very positive sort of way, and since no direct evidence can be found of any force or momentum driving mutations in a direction they have already begun, evolutionists entirely disregard the possibility that orthogenesis might be a cause of evolutionary change.

This brings me to an end of my preliminary discussion of theories of evolution which have rivalled the synthetic theory. We have seen that all three have failed to explain the basic facts of heredity, and this is the major reason why they are disregarded today. However, there are other reasons for their rejection (and conversely, for the acceptance of the synthetic theory). These reasons lie in the subsidiary areas of evolutionary study, and I shall now briefly consider three—morphology, systematics, and paleontology.

6.4   *The evidence of morphology*

Morphology is the science which tries to explain the physical characteristics of organisms—the eyes of men, the green leaves of plants, and the fins and scales of fish. Obviously, the major contribution of the synthetic theory to this area of study is the supposition that phenotypic organic characters are essentially adaptive—that is, that such characters exist and have the form that they have because by such existence and form, they help their possessors to survive and reproduce. As we have just seen, two rival theories, the saltation theory and the orthogenetic theory, would want to challenge at least a part of this assumption. What evidence is there to help decide this dispute?[3]

It seems clear that many organic characters are in fact adaptive. We have already had some evidence of this, namely the colour of moths; but one need carry out no elaborate tests to see adaptiveness around us. Eyes, teeth, hands, ears, legs—they all aid in survival and reproduction. Admittedly, an animal might loose a leg and still raise

[3] Here, incidentally, I think we have another example of an area of evolutionary studies which gives no real way of deciding between the synthetic theory and all of its rivals—I suspect that the Lamarckian's claims about morphology, and indeed about systematics and paleontology, would not be very different from those of the synthetic theorist, although were one to accept Lamarck's views about a movement towards perfection, pertinent differences might appear. But see Rudwick (1972) for Lamarck's attitude towards the fossil record.

a large family; but the chances are much weighted in favour of an animal with all of its legs. Moreover, against the saltation theory, it clearly seems the case that the basic 'plan' of organisms is adaptive. It might be an interesting experience if our eyes were on the small of our backs; but for organisms like us, walking in the way that we do, it is clearly much more useful to have them in front, up at the top of our bodies (and this, of course, is quite apart from the anatomical 'engineering' which would be needed to put our eyes on our backs). Hence, there is obvious truth in the claims of the synthetic theory.[4]

However, having admitted truth in the basic morphological claims of the synthetic theory, a big problem must be acknowledged. Much of our thinking about the existence and nature of particular adaptations is highly conjectural, or, at least, it was until very recently. For example, only fifteen years ago Dobzhansky went as far as to write of his own special field of interest that 'Indeed, most of the morphological traits which differentiate related species of *Drosophila* are of a kind the significance of which for the welfare of the animals is not at all evident' (Dobzhansky, 1956, 338). Speaking more generally, he admitted that 'It is a fact that the variation in some traits which at one time seemed neutral was later discovered to be adaptive' (Dobzhansky, 1956, 339). Yet, he concluded that 'it is still as true today as it was in Darwin's day that the adaptive significance of most traits which vary between and within species is obscure. Suffice it to mention that practically nothing is known about the significance of most traits which vary between races of man as well as among individuals within a race' (Dobzhansky, 1956, 339).

Obviously, much remains to be learnt about adaptive advantage. We know enough to be able to claim that adaptation is widespread; but we are far from being able to say (on the direct evidence) just how widespread it is, although I think it is fair to say that in the past couple of decades, primarily because of the brilliant field and laboratory work of such men as Dobzhansky (in America) and Ford (in England), the direct evidence in favour of the synthetic theory's suppositions about the widespread nature of adaptations has been dramatically increased. But even if we were to exclude much of this new evidence, before it could immediately be concluded that morphology fails to give major support to the synthetic theory, one import-

[4] Paradoxically, Goldschmidt, one of the strongest supporters of the saltation theory, argued that some differences between species are such sophisticated adaptations, they could not have come about by selection on small variations. 'The wing shape, type of flight, and correlated structure of the lungs of a hummingbird, together with a honey-sucking bill and tongue, are a beautiful adaptation to a definite ecological niche. But could selection favour such a combination before it has reached a working capacity?' (1952, 93). Goldschmidt's answer was 'No'!

ant point would have to be noted. Although the theory stresses the importance of adaptation, it is far from claiming that *all* characters must be adaptively advantageous. In fact, there are at least five reasons why evolutionists think some characters might not be adaptive (although not all evolutionists accept all of the reasons). Some of these reasons, we have already met. One is the possible existence of *genetic drift*, leading to the fixation of non-adaptive (or even positively disadvantageous) characters. Another is *pleiotropism*. If a pleiotropic gene causes two phenotypic characters, one of which is adaptive and the other inadaptive, it might be to the organism's benefit to tolerate the latter for the sake of the former. A third reason is *changed conditions*. A character may once have been adaptive, but, with changed environmental conditions, is no longer so. As yet, it has not disappeared from the population. A fourth reason is a character having *two roles*. The peacock's tail may attract the peahen; but at the same time, it may make its owner more vulnerable to attack. Fifthly, there is *allometric growth*. Some genes cause phenotypic characters to grow more rapidly than the rest of the body, and these characters continue growing after the sexual maturity is reached. They may give their owners the reproductive edge over other organisms, even though later they prove to be a handicap (by which time, the genetic damage is done). It is thought that the Irish elk's horns are possibly the size they are for this reason.

Perhaps by this point a critic of the synthetic theory might be feeling somewhat uncomfortable. With all these possible reasons for inadaptiveness, one might feel that there is no way in which a character could not fit the theory's claims. If a character is adaptive, then all well and good. If it is not, then it is not for one of the reasons given above. Hence, as von Bertalanffy argues: 'A lover of paradox could say that the main objection to selection theory is that it cannot be disproved' (von Bertalanffy, 1952, 89). There are, however, three replies one can make to this criticism. First, even if morphology could not disprove the theory, there are other sources of refutation, particularly genetics. Secondly, because evolutionists can give many reasons for inadaptiveness, this does not mean to say that if they cannot prove a character adaptive they can immediately claim it to be inadaptive. They must show why they think one of the reasons for inadaptiveness holds. A conjecture is not proof. Thirdly, none of the reasons given above is just an *ad hoc* device to save the face of evolutionary theory. Pleiotropism, for example, does really have independent evidence. Thus, evolutionists are not just conjuring up dodges to avoid awkward counter-examples.

Nevertheless, even if one grants all of these points, one might still feel some lack of comfort about the morphological evidence for the synthetic theory. One who feels this way is Rudwick, who, although he is specifically concerned with the paleontological evidence (for the synthetic theory), makes a criticism which is just as applicable to the morphological evidence. Rudwick writes as follows:

With a certain critical fossil organism before us, we may argue whether its distinctive features were or were not adaptive. But we cannot reach any conclusion about these features except by testing them as possible embodiments of conceivable operational principles. We may then be able to demonstrate that they were probably adaptations of high efficiency for particular functions; and by doing so we shall add cumulative weight to the synthetic theory (or any other theory that stresses the ubiquity of genuine adaptation). Yet if we are unable to do so, our failure will not add corresponding weight to some other theory, for it can always be said that the features may prove to have been adaptive, if only we can think of the right function. For there is no positive criterion by which non-adaptiveness can be recognized and demonstrated. At least in this respect, therefore, the synthetic theory would seem to be as unfalsifiable as its rivals are unverifiable. (Rudwick, 1964, 39–40)

Rudwick seems to find an asymmetry about the synthetic theory. Positive evidence counts for it, but negative evidence does not count against, and obviously if his argument is sound, then the asymmetry holds just as much in the case of the morphological (i.e. phenotypic) characters of living organisms as it does in the case of the morphological characters of fossils (although, this is not to deny that it could be much more difficult to carry out tests in the case of fossils). However, I would suggest that Rudwick's conclusion is not well taken, for his argument rests on a logical confusion. The point Rudwick overlooks is that whilst, admittedly, we can never be absolutely certain that some character is inadaptive, so we can never be absolutely (logically) certain that some character is adaptive. There comes a point when we assume that characters are adaptive; but, logically we could be mistaken. Since we are prepared to make this jump from probability to certainty in the case of adaptive characters, can we refuse to make the jump in the case of inadaptive characters? I would think not. In other words, although it is always (logically) possible that some apparently inadaptive character has an adaptive value, after a while it becomes unreasonable to assume that such an adaptive value does exist. Hence, contrary to Rudwick's claim, no such asymmetry exists between the morphological evidence for and against the synthetic theory.

Nevertheless, not unreasonably the critic might ask for some specific examples of kinds of morphological evidence which would count against the synthetic theory. What one needs to suppose is some kind of apparently inadaptive character which it would just not seem reasonable to put down to any normal genetical cause (like pleiotropism). As just one example, I suggest the following. Suppose one came across a group of mammals and that they all had one leg grossly swollen (as if caused by elephantiasis). Suppose that one carried out all kinds of tests on these animals and found that the leg does seem to be a handicap—the fatter the leg, the more liable they are to being caught by predators; the females, if anything, prefer thinner-legged males; there is no evidence of allometric growth; members of the same species can be found elsewhere, identical but for the leg (and hence pleiotropism seems implausible); and so on. Suppose, however, that one found no evidence that leg size was diminishing—on the contrary, it seemed to be increasing. Frankly, if one found something like this, I think that one would have to start to conclude that one has here an unexplained (and, within the context of the synthetic theory, inexplicable) inadaptive character. Of course, we do not think we shall find such a character, and it is worth noting in fairness that most unexplained characters are far from being so obviously inadapted as the example given above—most, like slight colour differences, seem neither very adaptive nor inadaptive—but the reason why we do not think we shall find such a character is not because evolutionists are constitutionally incapable of recognizing morphological counter-examples to their theory, but because, so far, no example like the one I have just described has occurred. If a theory has no counter-examples, it might possibly be because it is true. (On the topic of what kind of characters would actually falsify the synthetic theory, see the rather amusing exchange between Maynard Smith and Waddington in Waddington, 1969.)

6.5   Systematics

Systematics is, as we have seen, the study of the different kinds of animals and plants, and there is at least one good test from systematics relevant to the choice between the synthetic theory and the saltation theory. The synthetic theory begins, as it were, from the bottom and works up. Small variations add up to larger variations, and these in turn add up to still bigger variations. Hence, if the theory is true, one ought to find evidence of groups of organisms which are almost species, but not quite. That is, one ought to find groups which are on the verge of becoming reproductively isolated from each other. The saltation theory begins at the top and works down. First one gets really high macromutations, and then mutations of lesser and lesser effect. If this theory is true, speciation is a one-step affair, and hence, given any two populations of organisms, they ought to be unambiguously included in either the same species or in different species.

The expectations of the two theories clearly contradict each other, and it is the synthetic theory's expectations which are found to hold in the world. Some groups are obviously on the point of speciation (i.e. of forming separate species). Well-known instances of these are so-called 'rings of races'. Here one has a chain of groups, no group reproductively isolated from its neighbours, but with the end groups touching or overlapping each other, and remaining reproductively isolated. Were some of the central groups to be removed, these end groups would be counted two quite separate species. As it is, gene exchange is possible between them, via intermediaries. Hence, in one sense they are separate species, but in another sense, they are not. These rings of races are clearly species in the making. They could not exist were the saltation theory true. They are not only possible —they are expected given the truth of the synthetic theory.

6.6 *Paleontology*

Nearly everyone, if asked why they believe in evolution, would reply 'because of the fossils', and it is undeniable that as far as the *fact and course* of evolution is concerned, the fossil record is irreplaceable evidence. However, as far as the *theory* of evolution is concerned, evolutionists of all kinds have, following Darwin, seemed to spend far more time explaining why the fossil record does not refute their theory, rather than triumphantly forwarding it as evidence in favour of their theory. (The Special Creationists, of course, found the fossil record no more comforting.) Admittedly, since Darwin a great deal of fossil evidence has been uncovered; however, I think that most evolutionists would still want to agree with Carter, who writes as follows:

A theory of evolution can never be deduced from palaeontological fact alone. Fossils may show us the changes by which one animal has evolved from another, but they tell us very little of the means by which these changes were brought about. The palaeontologist may be able to exclude some theories of evolution on the ground that they demand change not in accord with his facts . . . But to build a theory of evolution is much more the concern of the biologists who deal with the nature of the animal organisms and with changes that can take place in it—geneticists, students of animal life histories, ecologists and others. The part of paleontology in the study of evolutionary theory resembles that of natural selection in the process of evolution; it serves to remove the inefficient but cannot itself initiate. (Carter, 1951, 9)

An important fact must therefore be recognized about the fossil record. It cannot give us any direct evidence of the mechanisms of evolution. It can only tell us about the effects of these mechanisms. But, as Carter points out, the record is a potential source of tests aimed at deciding between evolutionary theory and its rivals. Specifically, the record gives us tests between the synthetic theory and the saltation theory, and the synthetic theory and the orthogenetic theory. Let us look at these tests in turn.

At a first acquaintance the fossil record seems to support the saltation theory, since in most cases we do not find smooth evolutionary changes as if caused by very small variations. Instead, we find different kinds of organisms popping in and out of existence abruptly, just as one might expect if evolution were essentially a function of macromutations. However, it should be recognized that the synthetic theory does not claim that the fossil record must always show smooth evolutionary changes—what is claimed is that such changes must have occurred. The trouble is that the chances of something being fossilized, even given gradual changes, are fantastically small—particularly if it is something unlikely to fossilize anyway, like an insect. Just the right conditions must prevail for fossilization; then, if we are to get evidence of the evolution over a period, we need the same conditions holding for a period, with constant deposition. Moreover, from then on, the rock formations in the area must remain stable. Obviously, usually some of these conditions will break down, and hence, even if we do get evolution as supposed by the synthetic theory, we cannot expect much more of the fossil record than what we actually find.

Nevertheless, if the synthetic theory is indeed true, we should occasionally expect to find some intermediate organism linking one group to another. Conversely, if the saltation theory is true, such links could never be found. As the synthetic theory predicts, some such links are occasionally found. The most famous example is *Archaeopteryx*, an animal which is half-bird but still half-reptile. Such an organism is quite impossible given the truth of the saltation theory. Actually, we now know of many more, less dramatic links between groups of organisms. In fact, in the case of the transition from reptiles to mammals, there are now so many intermediate forms known that, during a relatively recent discussion, every participant chose a different point to mark the beginning of the mammals (Van Valen, 1964). Even the missing link between man and ape, so beloved of fundamentalist theologians, is now far less hypothetical than it was once (Mayr, 1963). Hence, despite its limitations, the fossil record gives little comfort to the saltation theory, and conversely, positive support to the synthetic theory.

In a like manner the fossil record shows that there is little reason to hold an orthogenetic theory, even though the record was one of the major reasons why such a theory was ever proposed in the first place. There are certainly evolutionary trends; but few are so obviously inadaptive as the orthogenetic theory would seem to demand. In many cases, in fact, the trends seem to be constantly in the direction of increased adaptive advantage. For instance, many animal lines show an increase in body size over a long period. However, within limits, increase in body size is adaptively advantageous. For example, the larger animal is often swifter and hence can catch food or escape from predators better. Moreover, as Simpson points out, were the synthetic theory true (as opposed to the orthogenetic theory) one would expect to find that these trends are far from regularly unidirectional. One would expect to find periods of stagnation and even reverse as the environmental conditions fluctuated. For example, in times of famine, small size might be adaptively advantageous over large size. It is precisely this irregularity which we often find in evolutionary trends. Indeed, sometimes there is so much irregularity it is a little pointless even to go on talking about trends. About the most famous supposed 'trend' of all, reduction in the side-toes of horses, Simpson writes:

There was no such trend in any line of Equidae. Instead there was a sequence of rather rapid transitions from one adaptive type of foot mechanism to another. Once established, each type fluctuated in the various lines or showed certain changes of proportion related to the sizes of the animals, but had no defined trend. (Simpson, 1953, 263)

Hence, it would seem that, if anything, the evidence points in the direction of the synthetic theory rather than the orthogenetic theory. Certainly the evidence for the latter does not seem anything like as good as was once claimed when the fossil record was less well known. However, it should not be forgotten that, as Carter points out, the main reason why modern evolutionists are neo-Darwinians (and neo-Mendelians) rather than orthogeneticists lies not in the fossil record. What really counts is that orthogenesis finds no support in other areas of evolutionary study, particularly genetics, whereas the synthetic theory does.

6.7   Conclusion

In the hope of finally stilling the critics let me conclude by restating the three major points about the evidence for the synthetic theory of evolution. First, one does not have to be able to make predictions about the future evolution of elephants before one can find any evidence for (and against) theories of evolution. The past as well as the present and future can give us evidence, and, given the terrific importance of genetics, although they are more prosaic animals, our knowledge of fruit-flies can be just as revealing as our knowledge of elephants. Secondly, evidence for a *theory* of evolution can be great, even though evidence for the actual *course* of evolution might be slight. (Note how, in the quotation on p. 98, Carlo speaks indifferently of 'evolution theory' and 'evolution'.) Thirdly, one must judge evolutionary theories by the evidence in *all* areas (but, most particularly, in genetics). Just to concentrate on paleontology, for example, is to get a very distorted view of the true picture. Of course, it cannot be denied that, as we saw, at least one critic of evolutionary theory, Himmelfarb, believes that to appeal to all of the evidence in this way is illegitimate. She feels that to do so is to rely on a 'logic of possibility' where, unlike conventional logic, possibilities add up to probability. However, I would suggest that Himmelfarb has the wrong analogy in mind here. She thinks of the process of theory confirmation as being akin to finding one set of probabilities given other sets of probabilities. Thus, if we have a 1 in 6 chance of throwing a six, then the chance of throwing two sixes is much less, namely 1 in 36. But, the situation we are considering is not like this. In our case, we have something very similar to the task faced by a prosecuting counsel when he wants to build a case out of circumstantial evidence. On its own, the fact that Lord Rake was shot by the butler and that the butler is a crackshot is not overwhelming evidence in favour of the butler's guilt. On its own, the fact that the butler's alibi breaks down is not overwhelming. On its own, the fact that Lord Rake seduced the butler's daughter is not overwhelming. However, add up all of these facts and others, and the existence of the butler's guilt is overwhelming. It is put 'beyond reasonable doubt'. In a like manner, because of all the evidence taken together, the truth of the synthetic theory (in the sense discussed at the beginning of the chapter) and the falsity of its rivals is beyond reasonable doubt.

7

TAXONOMY I

THE EVOLUTIONARY APPROACH

Given the incredible range of organisms on the earth, both living now and as revealed to us from the past through the fossil record, any biological theorizing obviously presupposes (or goes hand in hand with) an attempt at classification. This science of classification is called 'taxonomy' and in this chapter I shall be considering the traditional way of doing taxonomy—the taxonomy which, for obvious reasons, is called *'evolutionary* taxonomy'. In the next chapter I shall consider a newly arrived alternative to evolutionary taxonomy—the taxonomy which is known as *'phenetic* taxonomy' or *'numerical* taxonomy'. (The journals *Systematic Zoology* and *Taxon* contain many philosophical and quasi-philosophical discussions of the problems of taxonomy.)

7.1 *The Linnaean hierarchy and its evolutionary content*

A biological taxonomic system consists of a set of rules which enable one to classify in a particular way. The system which I am considering in this chapter owes its rules to the work of three men—Linnaeus, Darwin, and Mendel. Linnaeus provided what one might call the formal structure of the system; Darwin and Mendel set us on the road to the knowledge which enables one to give the Linnaean structure an empirical content. Let us take these two aspects of the evolutionary taxonomic system in turn, beginning with the Linnaean contribution (see Buck and Hull, 1966).

Following Linnaeus, organisms are classified hierarchically. One starts with a number of classes known as 'categories'. These in turn have as members classes called 'taxa'. Organisms are the members of taxa. Each organism is the member of one and only one taxon in each category. The categories themselves form an ordered set, and the position of a particular category relative to its fellows is called its 'rank' or 'level'. Each member of a category at a particular rank is included in a member of the category at the next higher rank. In other words, since the members of categories are taxa, given a particular taxon of a particular category, its members (which are, in turn, organisms) are also members of a taxon of the next higher-ranked category. This obviously applies all the way up the category rank, until one gets to the top-ranked category. The taxa of this category are included in no other taxa. Usually, the members of taxa at different ranks (a 'taxon rank' is that rank which a taxon's category has) are made increasingly more numerous as the rank is increased, so that one has many taxa in the lowest rank, and but a few taxa in the highest rank. (The way in which the members of taxa are made more numerous is, of course, by including more than one taxon of one rank in the same taxon at the next higher rank.)

A formal description of a hierarchy makes it sound much more complex than it really is. The following simple example will, no doubt, do more than a thousand words. Suppose one has ten organisms, grouped into four taxa at the lowest rank, and suppose also that there are three different categories. There is no one absolute way in which the grouping must continue up the hierarchy; but Figure 7.1 would be typical.

$$\gamma_3 \ \overset{G_3}{\Big[(a, \ b, \ c, \ d, \ e, \ f, \ g, \ h, \ i, \ j) \Big]}$$

$$\beta_2 \ \overset{E_2 \qquad\qquad F_2}{\Big[(a, \ b, \ c, \ d, \ e, \ f) \ (g, \ h, \ i, \ j) \Big]}$$

$$\alpha_1 \ \overset{A_1 \qquad B_1 \qquad C_1 \quad D_1}{\Big[(a, \ b, \ c) \ (d, \ e, \ f) \ (g, \ h) \ (i, \ j) \Big]}$$

Figure 7.1

In this example the small arabic letters stand for organisms, the curved brackets stand for taxa, the capital arabic letters are the names of the taxa (and the subscripts, the taxa's ranks), the square brackets stand for categories, and the Greek letters are the names of the categories (and the subscripts, the categories' ranks). As can be seen, each organism belongs to one and only one taxon in each category. Thus, for example, organism a belongs to taxon A at rank one, E at rank two, and G at rank three.

In the version of the Linnaean hierarchy which is used today, each organism must belong to a minimum of seven taxa which are members of seven specified categories of increasing rank. There are in fact more than seven categories, and these extra categories are used for more detailed classification; but belonging to a taxon of each of these seven categories is essential. As an example, suppose one attempts to classify a particular organism which belongs to one of the wolf species, then it automatically follows that there are seven taxa to which this wolf belongs and these taxa in turn belong to seven categories. The taxa and categories are as follows:

Category	Taxon
Kingdom	Animalia
Phylum	Chordata
Class	Mammalia
Order	Carnivora
Family	Canidae
Genus	*Canis*
Species	*Canis lupus*

As can be seen, convention demands that the name of the genus be included in the name of the species (both of which are italicized).

So far, we have been concerned only with the formal structure of the evolutionary taxonomic system. Let us now turn to the empirical content which supporters of the system try to put into it. Thanks to Darwin, Mendel and their successors, we now recognize that the organic world as we find it—both here and now, and in the fossil record—is the product of an evolutionary process. It is the aim of evolutionary taxonomists to reflect in their system something of this evolution of organisms; however, one thing must be recognized immediately. No one, not even the most ardent evolutionist, attempts to build into his system every iota of evolutionary information that he has concerning the organisms he is trying to classify. This means that the evolutionary taxonomist must make some decisions about exactly what kind of information he wants to put into his classification; and, not surprisingly since evolutionists do have different interests, this means that despite a common overall commitment to evolutionary ideals, one does find somewhat different approaches when it comes to more particular problems. Admittedly, as far as the actual practical business of classifying is concerned, these differences do not always have a great effect on the finished products of taxonomists; but their theoretical importance should not be underestimated.

Basically, there seem to be two major approaches to evolutionary taxonomy in the English-speaking world, although, as we shall see, there is quite a bit of overlap both in the aims and principles of these approaches. On the one hand, one has what one might call the *geneticists* (see Mayr, 1969). For them, the dominant consideration is not so much the past history of organisms (even though they do not entirely ignore it), but the similarities and differences of the genes and gene complexes of the organisms they are classifying. As Mayr, one of the leading supporters of this kind of taxonomy, writes: 'When an evolutionary taxonomist speaks of the relationship of various taxa, he is quite right in thinking in terms of genetic similarity, rather than in terms of genealogy' (Mayr, 1965). On the other side of the fence, one has what one might call the *genealogists* (see Simpson, 1961). For them, evolutionary classifications must, in some sense, reflect organisms' phylogenetic history. Thus, Simpson, one of their spokesmen, writes: 'It is preferable to consider evolutionary classification not as expressing phylogeny, not even as based on it (although in a sufficiently broad sense that is true), but as *consistent* with it. *A consistent evolutionary classification is one whose implications, drawn according to stated criteria of such classification, do not contradict the classifier's view as to the phylogeny of the group*' (Simpson, 1961, his italics).

When we come to see how these different intentions translate into practical terms, we find that both geneticists and genealogists agree about one thing, namely that the species calls for a definition different from those used for higher categories. Take first the geneticists. They commonly speak of the species concept applying to groups of organisms which have a common 'genotype', or which share in a common 'gene-pool'. Less metaphorically, Mayr has argued that the taxonomic species (i.e. the category occurring in the Linnaean hierarchy) should correspond to the 'biological' species, which is a concept applying to 'groups of actually or potentially interbreeding natural populations, which are reproductively isolated from other such groups' (Mayr, 1942, 120). In the case of categories higher than the species (say, rank n), we find that geneticists argue that these classes, having as members taxa of rank n, should include in their member-taxa one or more taxa of rank n-1 'of presumably common phylogenetic origin, separated by a decided gap from other similar groups. It is to be postulated for practical reasons, that the size of the gap shall be in inverse ratio to the size of the group' (Mayr, 1942, 282). It should be noted that by 'decided gap' in this context is meant something genetical—there is, as it were, a genetical gulf between taxa of the same rank.

Turning next to the definitions of species and higher categories offered by genealogists, we find that, in part, the definitions are couched in the same language as the definitions of the geneticists; but, as we shall see, the meaning is different. The category of species, for Simpson, has as members groups which are what he calls 'evolutionary species', where 'An evolutionary species is a lineage (an ancestral-descendant sequence of populations) evolving separately from others and with its own unitary evolutionary role and tendencies' (Simpson, 1961, 153). For the higher categories, Simpson uses Mayr's definition; but it is clear that when he thinks of a 'decided gap' between one taxon and another, he does not have the same thing in mind as Mayr. Obviously, both Mayr and Simpson must normally work initially primarily from the *morphological* differences between organisms (with perhaps some few additional behavioural and other differences thrown in); but whereas for Mayr, as we have just seen, these differences are taken to reflect *genetical* differences between organisms, for Simpson, any morphological difference (between the members of two taxa) is to be taken as indicative of *the length of time since the two taxa split apart* (i.e. of the time since the members of the two taxa could properly be put in a single taxon of the same rank as these two). Thus, for example, Simpson justifies his use of morphological criteria in classifying by writing that:

This procedure's consistency with phylogeny depends in the main on two well-established and now familiar reciprocal principles:

1. Characters in common tend to be proportional to recency of common ancestry. The distances between lower taxa in this approach are inversely proportional to characters in common.

2. Degrees of divergence tend to be proportional to remoteness of common ancestry. Sizes of gaps thus tend to be directly proportional to remoteness of common ancestry among surviving lower taxa. (Simpson, 1961, 192)

These, then, in their barest outlines, are the major kinds of ways in which English-speaking evolutionists try to give the Linnaean hierarchy an empirical content. For the rest of this chapter I want to consider some problems of philosophical importance stemming from evolutionary taxonomy. I shall begin by considering the category of species. Next, I shall look at problems concerning the higher categories. Finally, I shall discuss briefly a criticism which has been made of the formal structure of evolutionary taxonomy, namely the Linnaean hierarchy itself.[1]

7.2   *The category of species*

In this section I want to look at the category of species—in particular, I want to look at the so-called 'biological species', which, as we have seen, Mayr defines as being something whose members 'are groups of actually or potentially interbreeding natural populations, which

[1] The chief omission from my discussion, a function of limited space, will be of the so-called 'phylogenetic' school which is popular in Europe. A discussion of the school's views, together with many references, can be found in Hull (1970a).

are reproductively isolated from other such groups'. At the end of the section I shall try to link up my discussion with Simpson's 'evolutionary species'. (Some of the points in this section are treated in rather less detail, and some in rather more detail, in Ruse, 1969. But see also Hull, 1970b, and Ruse, 1971b.)

There seems to be common agreement amongst biologists, or at least, in the writings of biologists, that there is something rather special about the biological species, or rather, about any group which falls into it. Somehow, groups which are biological species are felt in some sense to be 'real' in a way that other groups are not felt to be. Groups which are higher taxa or groups fitting some other species concept are thought to be mere fictions. In particular, groups not thought to be real are those made using the so-called 'morphological species concept', where this is a concept based on the assumption that organisms can be grouped into taxa on the basis of some kind of 'overall' morphological similarity and difference. Morphological species, according to the concept, are the smallest collections of organisms possessing this kind of overall morphological similarity, separated from other organisms by a distinct morphological gap, and containing no such internal morphological gap separating the members of the species from each other. Hence, men are men because they have man-like characteristics, and are to be separated from cows because there is a morphological gulf between things having man-like characteristics and things having cow-like characteristics. Almost without exception, evolutionary taxonomists are adamant in their contention that it is biological species alone which are real—morphological species are fictions or unreal.

The immediate question which calls for answer is just why it is thought that the biological species concept applies to real groups, whereas other concepts, particularly the morphological species concept, do not. A great deal of ink has been spilt on this, the 'species problem', and in order to prepare the way for what I find to be a satisfactory solution, I shall first consider two other attempts at solving the problem (see also Mayr, 1957; Lehman, 1967).

Mayr (1963; 1969) argues that biological species ('biospecies' for short) are in some important sense *objective*, whereas morphological species ('morphospecies' for short) are *subjective*, and I think that many taxonomists would agree with him. Unfortunately, however, although Mayr is no exception to the rule that every taxonomist refers to his own system as 'objective' and to his rivals' systems as 'subjective', he is also no exception to the rule that no taxonomist seems prepared to explain to the reader in just what sense he is using the terms. Lacking any direction from taxonomists (and from Mayr in particular) I assume that by 'objective' is understood something lying outside of ourselves, public, and in some sense, repeatable by different people under specified circumstances. By 'subjective', I assume is understood something internal, private, and not necessarily subject to repetition by different people (under specified circumstances). I take the distance between the Earth and the Moon to be objective; I take the 'obscenity' of ink-blots to be subjective. Others might disagree; but, in the absence of correction, I would claim that this way of construing the dichotomy accords with common practice. I would add that I do not think that 'total objectivity' or 'total subjectivity' are, despite the claims of many taxonomists, ever completely achievable (or that indeed such extreme notions make a lot of sense). I shall have more to say about this point later.

Is Mayr correct in his claim that the objective–subjective distinction solves the species problem? I think not. Even if we grant that biospecies are basically objective in the sense construed above, it is by no means obvious to me that morphospecies are not similarly basically objective. Whether or not it is possible to quantify degrees of morphological similarity and difference is a question we shall consider in Chapter 8. However, even though quantification is often a mark of objectivity, it is by no means a necessary condition—sometimes we all just preanalytically recognize things as being one way rather than another. This seems to be the case in the instance of morphologically distinguished species. Human beings just seem to have the ability to distinguish between groups of organisms on the basis of morphological difference. Cows look like cows, and horses look like horses. Indeed, even Mayr himself has recognized this fact, because he has written that:

. . . the most primitive native people have names for kinds of birds, fishes, flowers, or trees. If only individuals existed, and the diversity of nature were continuous, it would be difficult to sort them into groups and distinguish 'kinds'. Fortunately, at least in the animal world, the diversity of nature is discontinuous, consisting in any local fauna of more or less well-defined 'kinds' of animals we call species . . . [Moreover], Primitive natives in the mountains of New Guinea will distinguish the same kinds of organisms as, quite independently, does the specialist in the big national museums. (Mayr, 1969, 23–4)

In a like manner, the authors of an important recent book on plant taxonomy write that '. . . . as has been pointed out on several occasions, there is a very broad measure of agreement amongst practising taxonomists about the limits of taxonomic species in well-studied areas . . .' (Davis and Heywood, 1963, 94—they call the morphospecies the 'taxonomic species'). Hence, if by 'objectivity' we mean something 'out there', recognizable independently by different people, then it seems hard to deny that morphospecies can be reasonably objective, or, at least, that a good proportion are. Consequently, Mayr's appeal to the objective–subjective dichotomy does not seem to solve the species problem.

One objection that Mayr might have to this conclusion is that I am using the term 'morphospecies' in the above argument in a rather broader sense than one which refers just to morphological characters. I seem to be arguing that one can 'objectively' delimit groups on the basis of morphology *together with* behaviour and any other relevant characters. Certainly he could point out that, for example, the New Guinean natives probably use bird-calls to help to distinguish different kinds of birds. In a limited way this objection is well-taken; although it is clear that even such broadly construed morphospecies would be considered 'subjective' by Mayr, so long as their delimitation was made without respect to genetic relationship. But from now on, the reader can interpret my use of the term 'morphospecies' in a broad sense, although I would point out that usually one must and can delimit groups just on morphology in a narrow sense.

Another approach to the species problem is that of Simpson. He suggests that we speak of biospecies as 'non-arbitrary' groups, whereas others, including morphologically defined groups, are to be spoken of as 'arbitrary'. Simpson writes:

I believe that most of the purely semantic confusion on the present subject can be avoided if such terms as 'real', 'natural', or 'objective', and opposite or contrasting terms, are not applied to taxonomic categories or methods of classification, and if the two terms 'arbitrary' and 'non-arbitrary' are used in specially defined senses. (Simpson, 1951, 286)

He continues:

I propose to call taxonomic procedures arbitrary when organisms are placed in separate groups although the information about them indicates essential continuity in respects pertinent to the definition being discussed, or when they are placed in a single group although essential discontinuity is indicated. Conversely, procedure is non-arbitrary when organisms are grouped together on the basis of pertinent, essential continuity and separated on the basis of pertinent, essential discontinuity. (Simpson, 1951, 286)

A good many taxonomists have taken up this proposal by Simpson and blithely assume that this gives them a way of choosing biospecies over morphospecies. Unfortunately, what they do not realize is that the use of 'arbitrary' and 'non-arbitrary' can occur only *after* one has made one's choice in favour of a particular basis for classification. One first decides on whether one wants to classify according to reproductive criteria or morphological criteria, and then, and only then, can one say whether the taxa in one's category are arbitrary or not arbitrary. Thus, if one makes one's standards interbreeding and reproductive isolation, then biospecies alone will be non-arbitrary groups (as opposed both to higher taxa which include biospecies and to taxa formed on morphological grounds). But the point at issue is why one should choose reproductive criteria in the first place—one might have chosen morphological criteria, in which case morphospecies would be the only ones which could be non-arbitrary. Hence, again we have no solution to the species problem.

In any case, there are several reasons showing why, even if one adopted Simpson's proposal, the arbitrary–non-arbitrary dichotomy does not leave matters entirely unsettled. This is because decisions, arbitrary in Simpson's sense, have to be made about the classification of organisms, despite a prior decision to use the biospecies concept. Asexual organisms are reproductively isolated from everything. Hence, any attempt to group them involves arbitrariness (conversely, leaving them ungrouped is not to classify at all). Rings of races, encountered in the last chapter, mean that one must be arbitrary with respect either to reproductive isolation or interbreeding. Most far-reaching of all is the fact that, as soon as one starts to consider any kind of time-dimension, one runs into theoretical difficulties involving arbitrary decisions. Admittedly, usually the paleontologist can delimit his species on breaks in the fossil record; but if claims about evolution are true, there has been continuous reproduction back to a few original forms. Since, apart from polyploids, one generation is not reproductively isolated from its parent generation, without some arbitrary decisions one has to lump everything together, which again is not to classify. Consequently, even given the biospecies concept, one must be arbitrary in Simpson's sense.

It would seem that neither Mayr nor Simpson can give us conclusive reasons for preferring the biospecies concept over the morphospecies concept. Nevertheless, I think there is a very good reason why evolutionists do feel as they do about biospecies—and why they feel uncomfortable with the idea of the morphospecies concept presented in isolation. However, I do not think that we can find this reason whilst we consider the two species concepts *apart from any theory*. Just contemplating a biospecies out of context does not show the desirability of the biospecies concept, although I think this is the only approach we have seen offered so far. To find the right answer to the species problem, let us begin by asking ourselves why other scientists—particularly physical scientists—consider that some concepts tell us something significant about reality, whereas others do not. Why, for example, does the physicist pick out the concept of 'work' (defined as force × distance) for special attention? Why not pick out some concept defined as $(force)^3 \times (distance)^4$? The answer, obviously, is that a concept like 'work' enters into laws

—thus, for example, as Joule showed there is a fundamental connexion between the work done to heat a liquid and the heat produced— whereas a concept like (force)^3 × (distance)^4 does not (except derivatively in so far as force and distance independently enter into laws).

However, there seems to be slightly more than this to the problem of which scientific concepts tell us about reality. Suppose that we have a metal which is called 'guelphite' and that a number of known laws hold of guelphite. For example, we might know that 'guelphite has a specific gravity of 10·2'. Now suppose someone proposed dividing guelphite into two kinds, guelphite occurring in pieces under 10 pounds and guelphite occurring in pieces over 10 pounds. We would not think that his division corresponded to anything 'real', even though he were to point out that 'Any piece of guelphite under 10 pounds has an S.G. of 10·2' seems nomically necessary. However, were someone to propose a division based on the fact that some guelphite melts at 1000° C and the rest at 1100°C, we would probably feel much less worried about this division. Why is this? I suggest that it is because past scientific experience shows that something like melting point is likely to be reflected in some other characters that the metal has (unlike size of piece), and that these other characters give us another way of distinguishing the two kinds of guelphite. Thus, we might expect low temperature guelphite to have one kind of atomic structure and high temperature guelphite to have another kind of atomic structure, and instead of talking about temperature, we could distinguish the kinds of guelphite on the basis of this structure. In other words, we would think the temperature division reflects reality because (unlike the size-division) we have an alternative, logically independent way of characterizing the division (i.e. in terms of atomic structure)—a way which is linked to the first way by laws.

Hence, putting together the parts of this discussion, I suggest that if we consider that a concept reflects reality, this is because this concept enters into laws in such a way that one gets an alternative, logically independent way of making the distinctions made by the concept. That is to say, we think that A tells us about reality if and only if all A's are (nomically necessarily) also B's and vice versa (see also Schlesinger, 1963).

Returning now to biology, let us see if this criterion of scientific reality, which, because I have done so in the past, I shall call 'Maxwell's criterion',[2] helps us at all with the species problem. I think it is

[2] More properly it might be called 'Whewell's criterion' because that much-underrated philosopher was fully aware that the thing which makes a classification 'real' or 'natural' is 'that the arrangement obtained from one set of characters coincides with the arrangement obtained from another set' (Whewell, 1840, 1, 521).

fairly clear that it does. Suppose we have a group of organisms which has members interbreeding between themselves but isolated reproductively from all other organisms. These reproductive habits must be indicative of the fact that the members of the group share a common 'gene-pool', whereby this is meant that they possess a fundamental genetic similarity between themselves—a similarity not shared by other organisms. (I think this is what is usually meant by the term 'gene-pool'—but the exact normal usage does not affect my point.) If these genetic relationships were not the case, then clearly, given that individual genotypes are such delicately balanced things, these individuals could not form new genotypes with other sharers of the gene-pool, nor would they as a group be protected from genetic contamination from outside. But, what can we say about such a group with a common gene-pool (on the basis of biological theory)? One thing that we would expect, given the gene-pool, is that the group would share a 'pool' of common morphological characters (and probably common behavioural characters also). Obviously, due to such things as balanced polymorphism, not every organism will have every character, but there will be some fairly basic kind of morphological similarity. This is, to a great extent, simply because since the genes are the units of function, genetic similarity usually implies morphological similarity. But in addition, if a character proves adaptively advantageous to one or a few sharers of the gene-pool, the genes responsible for the character will probably be passed on to other sharers (obviously of later generations), who in turn will develop the character. Hence, due to the fact that, despite exceptions, most of the characters that an organism has are a direct function of adaptive advantage, we are again led to expect that the sharers in a common gene-pool will also share in a common 'morphological-pool'. Furthermore, also through our knowledge of genetics, we would expect such a group sharing a common gene-pool to diverge morphologically from other groups, since successful new variation (brought about by mutation to new kinds of genes) will spread through the group, and probably such new variation (or such a combination of new variations) will be peculiar to the group. Other groups will not have exactly the same kind and combination of mutations, and even if they do have some mutations which are the same they will be subject to different environmental stresses and thus will probably have different adaptive needs from the first group—consequently, they will spread different variations through their groups.

But what all of this adds up to is that, given a group with a common gene-pool (i.e. a biospecies), the members will have a shared morphology peculiar to themselves—that is, they will be a morphospecies. (Mayr himself writes: 'The reproductive isolation of a biological species, the protection of its collective gene-pool against pollution by genes from other species, results in a discontinuity not only of the genotype of the species, but also of its morphology and other aspects of the phenotype produced by this genotype. This is the fact on which taxonomic practice is based' (1969, 28).) Does the connection also work the other way, from morphospecies to bio-species (as it must, if Maxwell's criterion is to be applicable)? I think it does, since if we have a morphospecies, we expect there to be some reason behind this, and genetical theory leads us to look for some kind of basic genetical similarity and to expect that the members of the group will be able to transmit the causes of their common variation between each other (through a number of generations), whilst, at the same time, the members of the group will be protected from contamination from outside. But this is to say that the theory expects some given morphospecies to be a biospecies. Hence, by applying Maxwell's criterion for scientific reality, since to say of something that it is a biospecies is also to let us know (by virtue of genetics) that it is a morphospecies (something which is logically although not causally independent) and vice versa, we can now see why biologists think that a biospecies is a 'real' grouping, in a way that any random grouping is not.

Also, by virtue of Maxwell's criterion, something rather puzzling about the morphospecies concept is made clear. Evolutionists strike the outsider (one at least) as being somewhat hypocritical in their attitude to the morphospecies concept. On the one hand, it is denied that mere morphology can lead to a real grouping; but on the other hand, in 90 per cent of the cases (in 100 per cent of the cases for the paleontologist) when evolutionary taxonomists try to distinguish species they have little more than morphology to go on anyway. Hence, their practice seems not to live up to their theory. However, now we can see that evolutionists are certainly right when they deny that morphology *alone* can lead to real groupings—it cannot, for it is only when morphology is linked to something else, namely genetic criteria and through these to reproductive links and gaps, that the morphospecies grouping becomes real. On the other hand, we can also see that, although evolutionists think the biospecies concept important, they can legitimately use morphological criteria in making their groupings. Although logically independent, the biospecies and morphospecies concepts are different sides of the same coin.

In fairness, I should add that I suspect that another factor in the biologist's thinking about species concepts has to do with our attitudes towards causes. Somehow, we frequently think that causes are more fundamental than effects. Thus, it might be claimed that the 'real' basis for the division of guelphite is molecular structure, which is the cause of other possible dividing criteria, like melting point. Analogously, it might be argued that reproductive criteria are more fundamental than morphological criteria because the former are the causes of the latter and not vice versa. Remember how Mayr writes that, 'The reproductive isolation of a biological species . . . results in a discontinuity of its morphology . . .' However, modifying this point (whatever truth it might have) is first the fact that the point in no way diminishes the importance of the satisfaction of Maxwell's criterion—if reproductive links and isolation do not lead to things like morphological similarities and differences, then (as we shall see shortly) their importance as methods of division vanish. Secondly, it is far from obvious that there is a straight cause and effect relationship between reproduction and morphology—in many respects, morphology seems just as much a cause of reproductive habits as an effect. Certainly animals can be attracted or repelled sexually by each other's looks. Thirdly and finally, the whole question of which is considered more fundamental for understanding in biology—cause or effect—is far from straightforward. This is a question which will be discussed in Chapter 9 and so no more need be said at this stage. Hence, whilst the causal position of biospecies criteria might play some role in biologist's thinking, I doubt if it is overwhelming.

Not unexpectedly, there are a number of objections which might be made to the solution to the species problem which I have offered. Rather than trying to anticipate them all, let me concentrate on what I think will be the major one; but let me do this by answering a criticism which is often made of the biospecies concept itself. A typical example of this criticism is that of Ehrlich and Holm (1962), who argue that the 'biological-species definition never has been operational and never will be'. The trouble is, they claim, that usually one has to work with morphological criteria, and then supposedly, one must infer reproductive facts, because it is impossible to tell directly whether most groups are in fact biological species. If groups are separated in time or space, then one cannot tell directly whether or not they are potentially interbreeding. And, argue Ehrlich and Holm, even if one brings members of two supposedly potentially interbreeding groups together artificially, then (as we saw in Chapter 6) one cannot be sure that they will perform as they would in the wild. Hence, since there 'seems to be an element of crystal-gazing in the idea of potential interbreeding' (since potentially inter-breeding groups have to be inferred on the basis of morphology), Ehrlich and Holm feel justified in rejecting the biospecies concept.

There are, I think, two points at issue here. First, Ehrlich's and Holm's general assumption that a biological concept must be 'operational'. Secondly, the problem with the biospecies concept itself. Let us take these points in turn—as we shall see, in discussing the second point, a major challenge to my solution to the species problem will be uncovered. The charge that some facet of biological theory is not 'operational' is one very commonly levelled by biologists; but in itself is not a very worrisome criticism. This is because operationalism as a prescription for science can easily be shown to be far too stringent for any sophisticated scientific activity. As formulated by its founder, Bridgman, operationalism demands that concepts in science be defined solely in terms of the operations involved in delimiting things falling under them. Thus, a ten foot length might be defined as the length covered by putting a ruler end to end ten times. To go beyond the operations is, we are told, 'not safe'. However, as many have pointed out, before we can get anywhere with science we must make moves the operationalist would judge 'not safe'. For example, if we use a different kind of ruler, unless we make an assumption over and above the operations involved, we cannot conclude that two objects measured by the different rulers could in fact have the same length. More generally, scientists must be allowed to make inductive generalizations in the course of their work, even though operationally such generalizations are 'not safe'. Thus, for instance, if a chemist talks of a substance being 'soluble', this does not mean that he has actually dissolved every instance of this substance (as the strict operationalist must demand). He has dissolved some instances, and on the basis of this feels justified in concluding that, given any instance, *were* it put in liquid, then it *would* dissolve. The making of generalizations like this is the foundation of the scientific activity. (Bridgman's thesis is discussed, with many references, in Hempel, 1954.)

The consequence we can draw is that, operational or not, the mere fact that evolutionary taxonomists use inductive generalizations to infer biological species from morphological species cannot be a cause for criticism. To do so, is to do no more than any other scientist (and to fail to do so, is to fail to do science). Hence, when evolutionists infer (on the basis of past experience) that some morphospecies is also a biospecies—arguing that the morphology of the group bears inferences about the reproductive features of the group—they are indeed doing things judged 'not safe' by the operationalist. However, what they are doing is relying on inductive generalizations (specifically, our knowledge of genetics in general and of other groups in particular), and the actual act of relying on such generalizations is to do no more than any other scientist does. It is a quite legitimate practice. (See also Hull, 1968, for devastating remarks on would-be biological operationalists.)

Nevertheless, this is not to say that the particular inductive generalizations involved, particularly that biospecies correspond to morphospecies and vice versa and that hence one can tell which groups would interbreed were they brought together, are legitimate tools of the scientist. Because of these claims, the critic might attack both the practice of evolutionary taxonomists and the solution I have offered to the species problem. If the correspondence does not hold, then everything falls apart—evolutionists can only rarely tell when groups are biospecies, and the nomic equivalence upon which my solution to the species problem depends is non-existent.

Now, it cannot be denied that the correspondence between biospecies and morphospecies does not hold exactly. Apart from the problem of asexual organisms, which organisms often form groups with remarkable morphological similarity between the members, there are two kinds of groups which show the correspondence not to be exact. The first kind are *sibling* species, where one has good species *qua* biospecies concept, but no morphological difference. The second kind are *polytypic* species, where one has groups which make good species *qua* morphospecies concept, but little or no reproductive isolation between the groups. These groups clearly make the various species concepts a little fuzzy around the edges, or, as Körner (1966) has put it, 'conceptually inexact'. The laws linking the concepts are obviously loose.

One cannot minimize the difficulties brought by the sibling and polytypic species; nevertheless, there are a couple of points which can be made showing that all is not lost, either for my solution to the species problem or in justification of the practice of evolutionary taxonomists. In the first place, specifically in defence of my solution to the species problem, if what I have written is true, then what we should find with such difficult cases (as sibling and polytypic species) is that evolutionary taxonomists compromise—sometimes dividing groups by reproductive criteria and sometimes by morphological criteria. If what someone like Mayr normally writes were true, then the evolutionary taxonomist would *always* decide in favour of the reproductive criteria. However, on at least one occasion, Mayr himself has written that 'where morphological and ecological differences are not discernible . . . it would seem impractical to separate these forms as species in routine taxonomic work'. Conversely, he argued that 'to combine all morphological species that freely hybridize in zones of contact also leads to absurdity' (Mayr, 1957, 376). Hence, even these difficult cases support my position, although they do show that the reality of species is not always entirely clean-cut.

The second point to be made is that, from a practical viewpoint, although sibling and polytypic species make the inference from morphology to reproduction hazardous, evolutionary taxonomists are not without some recourses when faced with such difficult kinds of species. For a start, morphological definitions today almost invariably involve what Wittgenstein called the idea of 'family resemblance'. A particular morphospecies definition will include many characters, all of which are possessed by some members of the group, but none of which are possessed by all members—each member just having some of these characters. Thus, by such definitions, which evolutionists call 'polythetic' definitions, evolutionists can acknowledge and cope with the great internal morphological diversity shown by many (interbreeding) groups—groups which nevertheless are still quite morphologically (as well as reproductively) separated from other groups. Secondly, by using their knowledge about particular kinds of organisms, evolutionists can invoke subsidiary inductive generalizations in order to infer the existence of biospecies in those parts of the animal and plant world where polytypic and sibling species are common. For example, polytypic species are common amongst birds, and so evolutionists would probably expect more internal morphological diversity of bird biospecies than they would of biospecies from most other areas (and presumably, they would adjust their theorizing accordingly). Sibling species are common in the insect world, and so, evolutionary taxonomists would probably expect less of a morphological gap between two reproductively isolated insect groups. Thirdly, to the reply to the criticism that the whole point of sibling species is that their members exhibit no morphological difference at all, it should be noted that Mayr (1963) argues that so far there is no case of two sibling species not showing some morphological differences when they have been studied extensively.

There is one final point that I want to make about the biospecies concept. I suspect that a lot of the worry about the notion of 'potential interbreeding' is, apart from the fact that it is misplaced, redundant. As we have seen, when biologists introduce the concept they talk not only of reproductive criteria but also use metaphors like 'gene-pool' and 'corporate genotype'. Now what I have tried to suggest is that to talk of a group sharing or having a common gene-pool is to say that the members share a fundamental genetic similarity, not possessed by other organisms. In other words, organisms with a common gene-pool form a kind of morphospecies in the genetic world. But, if this is so, then although biologists usually talk of reproductive habits and the possession of a gene-pool in the same breath, a biospecies defined in terms of reproductive criteria and a biospecies defined in terms of a gene-pool are not quite the same thing. The one definition is about the having or failing to have viable offspring, and the other is about the, connected but distinct, genetic nature of organisms. (Mayr seems to recognize this when he says that 'the reproductive isolation of a biological species . . . results in a discontinuity . . . of the genotype of the species . . .'.) But, if we do have these two definitions, by adopting the one in terms of gene-pools, that is in terms of genetic similarities and differences, not only can Maxwell's criterion still be satisfied, but since every organism has a genotype, even spatially separated organisms can be compared with an eye to the biospecies concept without any reference to 'potentiality'. (This is not to deny that genotypes must frequently be inferred or that they could be lost entirely in fossilized organisms. But they would exist or have existed, whereas this is not so and never has been so in the case of reproduction between isolated groups of organisms.) Of course, this suggestion of mine does depend on a decision to construe 'gene-pool' in the way that I have just done; but I do not think this to be implausible given Mayr's stated intention to classify by genetic criteria. In any case, however one uses the term 'gene-pool' the fundamental genetic similarities remain, and these could be the basis of a species concept, whether or not this would be acknowledged as the bio-species concept.

This concludes what I have to say directly about the much debated biological species concept. In order to end this section, let me briefly link up the biospecies with Simpson's evolutionary species. Simpson sees the major difference between a biospecies and his evolutionary species—a lineage (an ancestral-descendant sequence of populations) evolving separately from others and with its own unitary evolutionary role and tendencies—in that only the latter really has a time-dimension. A biospecies is, as it were, a temporal cross-section of an evolutionary species. Now, it is certainly true that the evolutionary species definition pays more attention to time rather than the bio-species definition, although I am not at all sure that a biospecies is totally without any time-dimension. Apart from anything else, even to talk of 'interbreeding' seems to make implicit reference to the time for one or two generations, and the same holds for 'reproductive isolation' (although perhaps the same is not true of talk of a 'gene-pool'). (See Cain, 1954, for a discussion of this point.) However, there is, I think, a bigger difference between a biospecies and an evolutionary species than Simpson recognizes. The definition of a biospecies (and also the definition of a morphospecies) seems to make no reference to any particular time, place, or thing. To say that an organism is a member of biospecies x is to say something of its ability to form reproductive links with the members of x.

Similarly, to say that an organism is a member of morphospecies y is to say something about morphology. Both species x and y could admit as members, organisms from outer-space, from Earth's prehistory, and organisms made by man. One does not even have to believe in evolution to hold x and y, and Mayr in fact points out that pre-Darwinians did actually delimit groups using a form of the biospecies concept. However, given an evolutionary species z which evolved here on Earth, it does seem to me that we have a reference to the Earth, and to a particular place and time. I do not see how an organism could be a member of z, unless it were actually born of another member of z here on Earth at the time and place where z flourished (or flourishes). If we allow outside organisms, then we no longer have an ancestral-descendant sequence of populations—they are not evolving separately from others, and they certainly do not have their own *unitary* role—the outside organism is not a descendant, if it evolved at all its evolution was separate, and its role was not at one with the members of z. (Of course, one might have an extra-terrestrial evolutionary species—but then, it would make implicit reference to the place where it evolved.)

The way in which the definition of an evolutionary species makes essential reference to events in the Earth's history would seem to have implications for the question of whether or not biological concepts have some kind of historical element (in a way that physical theories and concepts do not). As promised, I shall have something to say about the historical nature of biological concepts in Chapter 10. Here, I want merely to point out that even if one were to adopt Simpson's concept of an evolutionary species, none of my conclusions in earlier chapters of this book would be affected. In particular, my claim that evolutionary *theory* makes no references to particular times, places, or things is unaffected. I do not think that the theory itself makes any reference to any particular species—at best, we get references to particular kinds of species (as in Bergmann's Principle). Moreover, there is nothing in the theory which bars its application to Simpsonian evolutionary species on planets other than the Earth. There might not be any such species, and, if there are, it is logically possible that the theory might not hold for them; but this would not alter the fact that the theory as we have it *now* makes no essential reference to the Earth and its history. Hence, my earlier distinction between evolutionary theory and phylogenetic histories is untouched.

7.3 *The geneticist's treatment of the taxa of higher categories*

No one thinks that there are any other groups existing in nature as 'real' as species; but an attempt must still be made at dividing organisms into higher taxa. As we have seen, a geneticist like Mayr believes that the proper basis for such division is genetic resemblance and difference. In this section I shall consider this proposal. I start with a theoretical objection to the proposal, and then I go on to consider an objection based on the supposed impracticality of the proposal.

A critic of the whole theoretical notion of a classification based on genes is, again, Ehrlich, who writes that:

> . . . it has proven very difficult to develop a clear idea of what is meant by genetic differences. Ignoring non-nuclear inheritance, we are still faced with the question of how genomic similarity and difference would be evaluated . . . . Is there some absolute criterion of genetic resemblance? If such a criterion can be found . . . then will the most useful classifications be based on genetic resemblance? (Ehrlich, 1964, 114)

Ehrlich himself argues that we can make no sense of genetic similarity and difference, and basically his argument revolves around the fact that, in some cases, the actual genes possessed by two organisms (or groups of organisms) are quite different, but the morphological characters are quite similar (e.g. as in the case of sibling species). Are we to say that such organisms resemble each other genetically, or not? Obviously, in one sense they do (since the genes work in the same way), but in another sense they do not (since the actual genes are different). Conversely, Ehrlich argues that sometimes we have organisms whose genes work quite differently (leading to gross morphological differences), but where the actual (structural) genetic difference involved is quite slight. His conclusion is that in real-life, where we get extremely complex situations, difficulties like these make the whole notion of genetic similarity and difference quite worthless.

My particular feeling about this criticism is part of my general feeling about most of the criticisms of evolutionary taxonomy. There is certainly merit in what Ehrlich claims; but it is doubtful whether his criticism is anything like as far-reaching as he himself seems to think (indeed, if the notion of 'genetic resemblance' is entirely vacuous, then I fail to see how Ehrlich could claim that the value of a taxonomy based on it would not be that great). In so far as Ehrlich is directing his attention against the possibility of some overall, absolute, unequivocal scale of genetic difference (or similarity), his objection is well taken. Obviously there would be something very strange about quantifying the genetic differences between viruses as opposed to the genetic differences between man. How, for example, could one hope to come to a decision about which were the more alike—organisms with complex gene systems with more similarities and with more differences, or organisms with less complex systems with less similarities but also with less differences? On the other hand, it is doubtful whether anyone, even an evolutionary taxonomist, has ever claimed that there could be such an absolute scale of genetic resemblance, nor has anyone ever tried to frame taxonomic systems which depend on such a scale. Although their work is part of an overall whole, evolutionists usually restrict their work to rather limited branches of the organic world (e.g. mammals), and concentrate on trying to work out relative differences between the members of these groups. Because of this restriction usually they insist that one should not try to judge the size of gaps used in one area by the size of gaps considered necessary in another area. For this reason, the interest of evolutionary taxonomists in genetic resemblances and differences tends to be restricted to the resemblances and differences existing between limited groups.

Now, whether or not there exists such a thing as 'genetic resemblance' in this more limited sense seems to me to be a very different question from that of the existence of an overall criterion of genetic resemblance. If we are dealing with organisms of roughly comparable genetic complexity—types of primate for example—then we do not have the incredible gap that we have between something like viruses and men. Moreover, whilst one would certainly not want to minimize the sorts of problems to which Ehrlich draws attention, for most organisms a big change in the structure of the genotype is reflected in a change in the phenotype (and vice versa) and the phenotypic changes brought about by alterations in one genotype tend to have corresponding changes in the phenotypes of organisms with comparable alterations in their genotypes. Hence, whether one takes genetic structure or function as being the more basic, one is going to come to a fairly similar estimation of genetic resemblance. To take a somewhat extreme example, the members of species of *Drosophila* have genotypes which are more similar structurally to each other than they are to men's genotypes, and exactly the same applies when we consider the function of genes rather than their structure. (This is not to deny, as I have just pointed out, that it would be a fool's errand to attempt to quantify differences between species of *Drosophila* and species of primate all on the same scale.) Thus, in a case like this, Ehrlich's fears are unfounded. Frankly, I suspect that most cases are much more like this than the sorts of instances Ehrlich mentioned, and hence it would seem a pity to throw over the unproblematic cases because of a number of difficult cases.

Moreover, even in these more difficult cases, I think some progress can be made towards an understanding of 'genetic resemblance'.

Why, after all, should one not distinguish between genetic resemblance *qua* function and genetic resemblance *qua* structure? Explicitly noting that we are understanding resemblance in one of these two senses, some attempt can clearly be made to group organisms genetically. If, for example, two organisms have the same genes but for one or two exceptions, we can speak of them being genetically similar (*qua* structure). This holds even if these few exceptions cause major morphological or behavioural differences. For example, one gene might change the ability of an organism to digest a certain kind of food—this could lead to quite changed behaviour given the availability or non-availability of alternative foods. Yet it would be strange to say that a great genetic difference (*qua* structure) was involved. On the other hand, two organisms might share the same food; but if the number, kind, and order of their genes differed a lot, it would seem proper to talk of the organisms as being genetically different (*qua* structure). In other words, what I would suggest is that in a limited way one can make some sense of the concept of genetic difference and resemblance.

This conclusion I have drawn is not to deny that the points Ehrlich makes must be recognized (and, where possible, answered) by those who would use such a notion. How greatly, for example, do we rate a difference which leads to reproductive barriers (even though the morphological difference may be slight) against a difference which leads to great morphological differences (but, perhaps, no genetic barriers)? Do we allow genetic differences to be a function of the internal structure of the genes, or do we concern ourselves only with the products of the genes? What weight do we put on the order of the genes, as opposed to their kind? Nevertheless, despite these questions which have to be answered, given the fact that some rough sense can be made of the notion of genetic similarity and difference, and given also the fact that most cases do not seem to present the kinds of difficulties Ehrlich highlights (since, in most cases, similarities and differences in genetic structure and function coincide), I would suggest that a foundation for classification like Mayr's, that is, one based on genes, should not be completely ruled out on theoretical grounds.

Now, let us turn and look at a practical objection which might be levelled against a classification based on genes (although those who make it do not draw too nice a distinction between evolutionary taxonomists, and thus would think it also holds against anyone who would classify on the basis of phylogeny). Obviously, if one is trying to classify according to either the genetic background or the phylogenetic history of organisms, one must start with morphology (in the broad sense understood earlier), and what one will do is consider certain morphological characters possessed by organisms to be of greater taxonomic importance than others. One is going to feel that some characters give one a great deal of information about the genetics or history of organisms, and that other characters do not give one so much information. Consequently, in classifying organisms, one is going to give certain characters greater 'weight' than other characters.

Nearly all the critics of evolutionary taxonomy object to this process of weighting, and they argue that it is unwarranted, arbitrary, and even that it involves circular arguments. The following passage by leading advocates of phenetic taxonomy, Sokal and Sneath (1963), is typical of the kind of objection which is raised. They write:

It may be advantageous at this stage to outline an important logical fallacy underlying current taxonomic procedure. It is the self-reinforcing circular arguments used to establish categories, which on repeated application invest the latter with the appearance of possessing objective and definable reality. This type of reasoning is, of course, not restricted to taxonomy—but it is no less fallacious on that account. Let us illustrate this point. An investigator is faced with a group of similar species. He wishes to show relationships among the members of the group and is looking for characters which will subdivide it into several mutually exclusive taxa. A search for characters reveals that within a subgroup A certain characters appear constant, while varying in an uncorrelated manner in other subgroups. Hence, a taxon A is described and defined on the basis of this character complex, say X. It is assumed that taxon A is a monophyletic or a 'natural' taxon. Thus every member of A (both known and unknown forms) is expected to possess X; conversely, possession of the character complex X defines A.

Henceforth group A, as defined by X, assumes a degree of permanence and reality quite out of keeping with the tentative basis on which it was established. Subsequently studied species are compared with A to establish their affinities; they may be within A, close to it, or far from it. It is quite possible that a species not showing X would be excluded from A, although it was closer overall to some of the members of A than some were to each other. It may be said that such problems would arise only when A was an 'artificial' group erected on the basis of 'unsuitable' characters. However, except in long-established taxa or those separated by very wide gaps from their closest relatives, the effect of the last classification carried out with a limited number of characters is quite pervasive. The circular reasoning arises from the fact that new characters, instead of being evaluated on their own merits, are inevitably prejudiced by the prior erection of taxon A based on other characters (X). Such a prejudgment ignores the fact that the existence of A as a natural (or 'monophyletic') group defined by character complex X has been *assumed but not demonstrated*. (Sokal and Sneath, 1963, 6–7)

As Hull (1967) has pointed out, if this passage does fairly describe the practices of evolutionary taxonomists, then their sin is not so much circularity as inconsistency. Supposedly, evolutionists consider widespread morphological similarities to be indicative of important relationships (and thus they justify the taking of character X seriously), and then, supposedly, they turn round and deny that widespread morphological similarities are indicative of important relationships (and thus they refuse to take seriously any widespread similarities which conflict with a classification based on X). Clearly, Sokal and Sneath cannot have their cake and eat it. If evolutionary taxonomic practice is circular, then it cannot be inconsistent, which latter is just what the above-described practice is. On the other hand, if the practice is inconsistent, then it cannot be circular.

Of course, none of this is to deny that, circular or inconsistent, if the above passage does truly describe evolutionary taxonomic practice, then there is something very seriously amiss with it. Classification is being applied almost entirely from above, without any regard for the real (evolutionary) nature of organisms. The question we must consider, therefore, is whether or not Sokal and Sneath are justified in condemning a system of classification which uses weighting. The answer, I think, is very similar to our earlier answers. No doubt, some classifications have risen little above the level of classification described by Sokal and Sneath above. However, having due regard for the purposes and aims of evolutionary taxonomy—to classify in the light of phylogenetic history or genetic background—justification can be found for the practice of weighting, that is, the practice of considering certain characters to be of greater value than others in the search for significant evolutionary relationships. To deny this is to deny evolutionists the virtues of inductive logic—virtues which, as we have seen, are open to other scientists. Nevertheless, even though weighting is not ruled out *a priori*, it will be seen that it still brings great difficulties—difficulties which the evolutionary taxonomist must try to overcome.

To justify my assessment of Sokal and Sneath's argument, let us look first at what Mayr writes in defence of weighting. He writes:

The scientific basis of a posteriori weighting is not entirely clear, but difference in weight somehow results from the complexity of the relationship between genotype and phenotype. Characters which appear to be the product of a major and deeply integrated portion of the genotype have a high information content concerning other characters (which are also products of this genotype) and are thus taxonomically important. Other kinds of characters, . . . as well as superficial similarities, convergences, and narrow adaptations, have low information contents concerning the remainder of the genotype and are thus of low value in the construction of a classification. (Mayr, 1969, 218)

What kinds of characters does Mayr believe to be of high weight, and conversely, what kinds of characters does he believe to be of low weight? Included in those of high weight Mayr mentions complex characters, jointly possessed derived characters, and characters which do not serve a specific, *ad hoc* adaptation. He justifies the choice of these as follows. First, the more complex a character, the less likely it is that two organisms having such a character have it by chance or that different gene complexes could produce such a character. The organisms probably have the same gene complexes and are closely related. Secondly, jointly possessed derived characters (i.e. characters which are the same but which have been acquired by two groups independently) point to the fact that the different gene complexes are reacting to environmental stresses in the same way. Because so many different responses are open in the face of stress, organisms would be unlikely to do this unless they were fundamentally similar genetically. Finally, characters which do not serve a specific *ad hoc* adaptation are not brought about by relatively rapid changes in the face of stress (that is, changes which are likely to involve little genetic change), 'but are merely, so to speak, an indication of an underlying basic genetic similarity' (Mayr, 1969, 220). The reasoning seems to be that if there is no real (i.e. adaptive) reason why characters should be the same, the chances are that it is because of genetic similarity. Characters Mayr puts at low weight include characters which vary a great deal amongst organisms, characters which are regressive (i.e. characters which involve loss, like the loss of teeth), and monogenic characters (i.e. characters which involve the change of only one gene). Mayr justifies his reason for so choosing them in a way very similar to his justification of the choice of high weight characters. For example, monogenic characters like albinism are obviously of low weight, because by definition they involve only minimal amounts of genetic change.

Now, as before, in order to evaluate Mayr's proposals let us note immediately that in appealing to a few known cases and applying them generally to support the giving of high weight to certain characters and low weight to other characters, Mayr is doing no more than any scientist must do, that is, he is relying on inductive generalizations. Moreover, let us also note that there can be no objection to his appealing for support to certain known facts about genetics. Thus, for example, if one were classifying certain organisms, say birds, one might decide to group members of species *a* with members of species in genus *b* rather than with members of species in genus *c*. One's reasons for so doing might include the fact that major differences between members of *a* and members of *b* involve colour, which past (bird) experience might have shown to be controlled normally by but a few genes, whereas major differences between members of *a* and members of *c* involve beak-shape which past (bird) experience might have shown to be a function normally of a very complex gene system. In such a case, because of one's past experience one would be giving greater weight to beaks and their changes in shape than to colour and its changes, and this certainly seems legitimate given the geneticist's aims. Hence, as in the case of the previous criticisms against evolutionary taxonomy, it would seem that a blanket condemnation of the practice of weighting is unmerited. One does not have to do a complete study on the genetic features of every organism in order to do a genetic classification; one can rather, apply the results which have been found to hold for a few. On the other hand, it should be noted that the practice of weighting is not always that straightforward. For a start, even if we admit the legitimacy of weighting, it is sometimes difficult to put it into practice. For example, one may be prepared to grant that the joint possession of derived characters is something which deserves high weight—however, one has first to decide which characters are really jointly derived and which are merely ancestral traits which have not yet been lost in either group. In the absence of a fossil record, to make a decision like this is, to say the least, not easy. Secondly, even if one can decide which characters deserve to be weighted highly and which do not, one has still got the problem of deciding how highly one should weight the various characters. For instance, does a complex structure rate more highly than a character that does not serve a specific *ad hoc* adaptation? As can be imagined, the phenetic taxonomists have seized upon this point, and, considering it insoluble, use it as an argument against evolutionary taxonomy. For example, Sokal and Sneath write:

If we admit differential weighting, we must give exact rules for estimating it. We must know whether the weight to be given to the possession of feathers is twice or twenty or two hundred times that given to possession of claws, and why. We do not know of any method for estimating this, and even if such a method were to be developed we doubt if any systematist would have the patience to use it because of the hundreds of characters he would need. (Sokal and Sneath, 1963, 119)

The pheneticists certainly have a good point here; but I doubt if the picture is quite as bad as Sokal and Sneath paint it. In the first place, by a system of trial and error, it would seem that evolutionary taxonomists could come to some rough judgments about the relative importance of different characters and combinations of characters. Admittedly, evolutionists do not normally try to quantify the different weights of different characters; but there seems to be no theoretical bar to their trying to do so. Moreover, pheneticists can hardly object to the very attempt to assign numerical values to different characters, because, as we shall see shortly, this is just what they try to do themselves. Secondly, Sokal and Sneath's dig about a systematist's not having the patience to use such a method of quantifying weighted character differences seems a little unfair. *A priori*, there seems neither reason why the evolutionist should need more characters than the pheneticist, nor reason why the evolutionist should have less patience than the pheneticist. Of course, none of this is to deny that any method of weighting, even one which sets up formal rules for assigning numerical values to different characters, would probably still leave a great deal of freedom to the individual classifier. However, as we shall see immediately, this subjective aspect of evolutionary taxonomy is not necessarily so dreadful a thing as the critics of evolutionary taxonomy suppose. Consequently, I would suggest that although the geneticists' approach to the classification of organisms into taxa higher than species is certainly not without obstacles, no insurmountable difficulties have yet been shown by its critics.

7.4   *The genealogist's approach to taxa of higher categories*

We have seen that Mayr claims that the evolutionists' use of the bio-species concept is made for the sake of 'objectivity'. Nevertheless, it still remains true that the most common criticism of all of the variants of evolutionary taxonomy is that it is 'subjective' (i.e. it lets different taxonomists come up with different classifications)—a situation which, the critics tell us, is quite untenable since 'classification must be freed from the inevitable individual biases of the conventional practitioner of taxonomy' (Sokal and Sneath, 1963, 49). Although this criticism is made against all evolutionary classifications, it becomes particularly strident in the case of the genealogist's classification of organisms into higher taxa, the approach favoured by Simpson. I shall, therefore, start this section by looking at the problem of subjectivity in evolutionary taxonomy, where this is understood as being the problem of what stems from the freedom of choice evolutionary taxonomy leaves for the individual taxonomist. Then, I shall consider an objection directed specifically against the practical possibility of giving a genealogical classification.

That the evolutionary classification of organisms into the taxa of higher categories does leave a great deal of room for the 'subjective' decisions of the individual taxonomist (i.e. decisions which call for a greater or less element of personal selectivity by the taxonomist himself) cannot be denied. This is particularly true of the genealogist's classification. Consider his rule to separate taxa from other taxa (of the same rank) by a 'decided gap'—not only must he himself decide on how big a gap a 'decided gap' must be, but also he must decide on whether or not great emphasis is to be put on the need to classify contemporaneous organisms together, or whether classification is always to be done 'vertically', that is, by putting together descendants and ancestors in the same taxon. For example, given the phylogeny in Figure 7.2 of the five species *A–E*, either of the two divisions into genera could be acceptable to the genealogist, and he must himself decide between them.

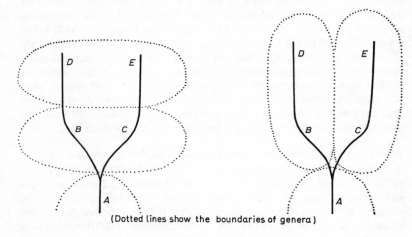

(Dotted lines show the boundaries of genera)

Figure 7.2

Of course, evolutionary taxonomists are not unaware of this kind of subjective nature of their classifications. As we saw, Simpson was at pains to stress that evolutionary taxonomies aim only to be 'consistent' with phylogenies—they do not aim to be unique reflections. Moreover, Simpson constantly talks of taxonomy being as much an 'art' as a 'science'. He writes:

Taxonomy is a science, but its application to classification involves a great deal of human contrivance and ingenuity, in short, of art. In this art there is leeway for personal taste, even foibles, but there are also canons that help to make some classifications better, more meaningful, more useful than others. (Simpson, 1961, 107)

The question which needs answering, therefore, is whether the great freedom that the evolutionary taxonomist (particularly the genealogist) has in making his classifications means that (as the critics claim) his finished results must necessarily be of low value. Is it the case that because evolutionary taxonomy lacks high objectivity and repeatability (and hence we can and do get alternative classifications from the same material), it is therefore something which should be replaced at the first possible instance? There are at least three points which show that, despite the great freedom it gives to the individual classifier, evolutionary taxonomy may still have some worth. In the first place, one must be wary of demanding too high a degree of objectivity or repeatability of any science. To condemn evolutionary taxonomy solely on the ground that it contains subjective elements is too hasty, because as we shall see later, the same can also be said of evolutionary taxonomy's rival, phenetic taxonomy. Indeed, as I have suggested earlier, the totally objective science of any kind—the science which eliminates all trace or responsibility of the scientist and which could not have been other than as it is—is a chimera. All sciences involve the paying of attention to certain facets of experience and the ignoring of other facets. They involve the need to idealize certain things in one way, and other things in other ways. They require, in short, the decision of the scientist to do one thing, when he might legitimately have done something else. If nothing else, the scientist nearly always chooses the simplest hypothesis which can explain the facts, and this decision is clearly 'subjective' in the sense that it rests on more than just something given, 'out there'. Of course, this does not hide the fact that evolutionary taxonomy might make many more demands on the scientist's personal judgment than most sciences; but to condemn evolutionary taxonomy just because it is subjective in places (in that it allows alternative classifications) is akin to condemning it because it is not operational (indeed, the mistake is probably the same, since operationalism invariably fights the Holy War under the banner of 'objectivity').

The second point to be made in defence of the classification (of higher taxa) produced by evolutionary taxonomists is that, despite the vagueness in the rules they offer as articulations of their practices, there does seem to be surprising agreement amongst evolutionists on how these rules ought to be applied (in particular, agreement on whether or not certain applications of these rules are good ones). We shall discuss shortly difficulties associated with the practical applications of the evolutionists' (particularly the genealogists') rubrics for classification at levels higher than the species; but let us note here that 'decided gaps' do seem to be very common in nature, even though evolutionists have difficulty explicating the exact meaning of 'decided gap'. Thus, one should not let the language of evolutionists seduce one into thinking that their discipline gives taxonomists more freedom than it really does.

Thirdly and finally, let us note that the freedom given to the evolutionary taxonomist is not necessarily a bad thing. As Simpson hints in the last quotation, different evolutionary taxonomists might be interested in stressing different things. The neontologist, for example, will be concerned primarily to show similarities and differences between living organisms. The paleontologist, on the other hand, has more interest in expressing relationships between organisms from different times. As things stand, both kinds of biologist have been left to pursue their own particular interests, and to provide their alternative classifications. No one (evolutionary) way of classifying is absolutely right, no one way is absolutely wrong —rather, each is useful in its own way. Hence, as I have just suggested, the subjectivity at the heart of evolutionary taxonomy— the freedom it gives to the taxonomist—has virtues, even if it has problems.

I come now to the final empirical question I shall raise about evolutionary taxonomy. If, like Simpson, one would classify in a genealogical way, then one must pay attention to the phylogenies of organisms. Several writers have objected to any rule of classification which presupposes some knowledge of phylogeny, the following criticism by Sokal and Sneath being typical:

The difficulty with the use of a phylogenetic approach in systematics emerged after the first wave of enthusiasm for it had subsided and has remained apparent to perceptive observers ever since. *We cannot make use of phylogeny for classification, since in the vast majority of cases phylogenies are unknown.* (Sokal and Sneath, 1963, 21; their italics)

The criticism is similar in many respects to the criticism by Ehrlich and Holm of the biological species definition—it is similar also in that it has the strength and the weakness of the Ehrlich and Holm criticism. It will be remembered that Ehrlich and Holm objected even to the use of inductive generalizations. In this criticism above, Sokal and Sneath seem to have a similar objection, and for this reason should be similarly faulted. Sokal and Sneath object that if we do not know a particular phylogeny, then we cannot use it for classification; but they give no reason why, *in principle*, one should not argue from known phylogenies to unknown phylogenies. This, of course, is just what someone like Simpson would want to do when he employs his assumptions that the number of characters in common tend to be proportional to recency of common ancestry, and conversely, degrees of divergence tend to be proportional to remoteness of common ancestry. He is arguing from known phylogenies where these propositions hold true, to unknown phylogenies where he thinks they hold true. The very act of using generalizations like these is no more to be condemned in the evolutionary taxonomist than it is in the physicist or chemist.

However, having said this, Sokal and Sneath are certainly right if they claim that the particular generalizations used by the genealogist are often in practice open to question. Just as sibling and polytypic species give difficulty at the species level, so also groups whose members change rapidly or slowly show that generalizations such as those used by Simpson are not entirely trustworthy. Yet, the genealogist like Simpson is probably right in claiming that, on average, change in character is proportional to length of evolutionary time, and, as in the case of species, the genealogist can invoke special hypotheses to deal with particular cases. If, for example, he knows (from known phylogenies) that some organisms are evolving (or have evolved) particularly rapidly or slowly, he might legitimately argue that the same special rates hold for other organisms with unknown phylogenies but which are similar in certain pertinent respects to those with known phylogenies. For example, suppose that one knew from phylogenetic evidence that a type of adaptation in mammals can be developed rapidly in the face of certain drastic environmental changes. Suppose also that one came across a group of mammals with such adaptations in an environment known recently to have changed in the relevant respects. Even though the phylogeny of this particular group might not be known, it seems (other things being equal) legitimate for the taxonomist to infer that the evolution of the adaptations in this group was rapid, and for him to use such information in his classifications. And he could properly say that his classifications of this group were reflections of phylogenies—even though the only direct information he had was morphology (and indeed, even though it was only morphology which led him to suppose that he was dealing with mammals in the first place). Hence, as in the case of most criticisms levelled against evolutionary taxonomy, there is some truth in the critics' claims, but not so much as the critics think.

7.5   *Gregg's paradox*

So far in this chapter, my sole concern has been with the nature of the empirical content evolutionary taxonomists try to put into their classifications. I want now to turn to a criticism directed more against the actual structure of the Linnaean system—although, as we shall see, the empirical content of the system is not entirely irrelevant.

A thing which one sometimes finds in evolutionary classifications is a *monotypic* taxon. This is a taxon which includes only one taxon from the next lower level. The best known case of monotypic classification involves the aardvarks, whose order Tubulidentata includes only one species, so that the order Tubulidentata, family Orycteropodidae and genus *Orycteropus* are all monotypic, containing just the same members as the species *Orycteropus afer*. Evolutionists use monotypic taxa so that they can show that things like the aardvarks are, from an evolutionary viewpoint, very far removed from all other living organisms. However, a number of set theoreticians, starting with Gregg (1954), have objected that monotypic taxa violate one of the major axioms of set theory, namely the *axiom of extensionality*. This states that classes having the same members are the same class. But, evolutionists most assuredly do not want to claim that the order Tubulindentata and species *Orycteropus afer* are the same class, even though they contain the same members. One is a taxon of the category order and the other a taxon of the category species, and basic to the Linnaean system is the premise that one taxon cannot belong to two categories (and similarly, one organism cannot belong to two taxa of the same category). Thus we have an impasse known as 'Gregg's paradox'.

There seem to be two ways out of the paradox. Either one makes taxonomy fit set theory, or one makes set theory fit taxonomy. Several people have tried taking the first course. Sklar (1964) suggests that each Linnaean taxon be replaced by a class containing all of the organisms of the taxon plus a number of the members of an arbitrarily chosen set of objects $a_1, \ldots, a_n$ (the a's are not to be organisms). Each new taxon of rank i will contain $a_1, \ldots, a_i$. Hence, in each case where we have a Linnaean taxon (of rank i) included in a Linnaean taxon (of rank $i + 1$), we have a similar relationship for the new taxa. However, no new taxon of rank i can contain the same members as the taxon of rank $i + 1$, because the taxon of rank $i + 1$ will contain a_{i+1}, whereas the taxon of rank i will not. Hence the difficulty over monotypic classification goes.

The trouble with this solution is that, as Gregg (1967) himself admits, we no longer have a system very acceptable to taxonomists. Taxonomists' taxa do not contain arbitrarily chosen a's. Gregg suggests a solution where monotypic taxa always contain one more organism than their included taxa; but although, as has been pointed out, if evolutionary theory is true, all monotypic taxa would cease to be monotypic if we could discover all of the organisms which ever existed (the aardvarks, for example, would have ancestral species included in the order Tubulidentata), none of these solutions at the expense of evolutionary taxonomy seem very satisfactory. The whole point of evolutionary taxonomy is that taxa are more than just collections of things—the only way that an extensional set theory can treat them. Moreover, it is not as if evolutionists want to use their taxa to make extensive formal deductions—thus, there seem to be no great benefits to be gained from altering taxonomy for the sake of the set-theoretic power conferred by the adoption of the axiom of extensionality.

I would suggest, therefore, that a satisfactory solution to Gregg's paradox must be one which pays attention explicitly to the nature and achievements of evolutionary taxonomy. In particular, as far as taxa are concerned, the axiom of extensionality must be dropped and it must be allowed that taxa of different rank (which are therefore different taxa) can have the same members. One can do this by permitting taxa names to have *intensional* definitions, rather than, as set theory requires, extensional definitions. This means that taxa names can be defined by specifying a number of properties required for taxon-membership, rather than by mere enumeration of the members. But intensional definitions are, of course, just what evolutionary taxonomists in fact use. Instead of stating that the taxon *Homo sapiens* consists of Michael Ruse, and John Gregg, and Pierre Trudeau, and etc., etc., they specify a number of properties for taxon membership. Thus, *Homo sapiens* is hairless, two-legged, big brained, and so on (actually, as mentioned earlier, evolutionists would probably make no one of these properties on its own absolutely essential for taxon-membership). If evolutionary taxonomists were not to use intensional definitions, then they could hardly have even the simplest taxon (especially if they allowed a potentially infinitely large taxon-membership) for, given any taxon, entirely new decisions would have to be made about every object one encountered in order to find whether or not it was a member.

But, if now we allow that taxa names can be intensionally defined, then Gregg's paradox is no longer troublesome. Although a monotypic taxon will have no more members than its sole included taxon, the two taxa will be kept separate by the fact that the taxa will have different requirements for membership. Thus, for example, suppose we have a taxon of rank $n + 1$, with a sole included taxon of rank n. The taxon of rank $n + 1$ might have a membership requirement of $A_1, \ldots, A_i$. The taxon of rank n will demand all of these A's for membership *and some more*. Hence, it will have a membership requirement of $A_1, \ldots, A_i, A_j, \ldots, A_m$, and it is logically possible that there could be other organisms, with properties $A_1, \ldots, A_i$, but without properties $A_j, \ldots, A_m$. Consequently, Gregg's paradox is paradoxical no longer, since although the taxa of different rank have the same members, the different requirements for taxon-membership keep them apart (see also Buck and Hull, 1966; 1969; Gregg, 1968; Ruse, 1971c).

8

TAXONOMY II

THE PHENETIC CHALLENGE

Whether or not, as one enthusiastic supporter has claimed (Ghiselin, 1966), evolutionary taxonomy 'epitomizes all that is *significant* in phylogeny', it cannot be denied that it is not without its rough edges. For some critics of evolutionary taxonomy the faults strike too deeply for them to feel that it could ever be in hope of salvation, and consequently these critics have devised a system of taxonomy of their own. I must confess that I am not always entirely clear about the major aims of this new 'phenetic' taxonomy. Occasionally, the primary aim seems to be to produce a taxonomic system which is 'objective' and 'repeatable' (i.e. a system which takes the decisions out of the hands of the individual classifier), and which leads to lots of good predictions. At other times, the primary aim seems to be to produce a taxonomic system which in some sense measures overall physical resemblance (called, by its practitioners, 'phenetic' re-semblance, and to be distinguished from 'genetic' resemblance). But whatever the main force of the replacement of evolutionary taxon-omy is intended to be, one invariably finds that the new taxonomists advocate a taxonomy which in some sense supposedly incorporates both aims. That is, it is supposed to be objective, repeatable, pregnant with predictions, and, in some way, based on phenetic resemblance. For the phenetic taxonomists, therefore, the phylo-genetic history of organisms is irrelevant, as is their genetic make-up. In this chapter I shall begin by considering the most important aspects of phenetic taxonomy—essentially as it is presented in Sokal and Sneath's *Principles of Numerical Taxonomy*. Next, I shall consider the evolutionary taxonomists' reactions to it. Finally I shall make some brief general remarks about the nature and aims of taxonomic systems.

8.1 *Phenetic taxonomy*

The basic principles and methods of phenetic taxonomy are fairly simple to understand. One starts with a number of organisms to be classified. These organisms, which are known as 'operational taxonomic units' (OTUs for short), are each deemed to be analysable into many different characters (40–60 at a minimum); and the similarities and differences between the organisms is assumed to be a direct function of the similarities and differences between their characters. A matrix is drawn up showing the various similarities and differences between the organisms (as calculated from the similarities and differences between their characters), and then various techniques are used to reveal and summarize the structure of the matrix. Usually the techniques used are numerical and are often collectively called 'cluster analysis'. As we shall see, there are different types of cluster analysis, but 'the important common aspect of all these methods is that they permit the delimitation of taxonomic groups in an *entirely objective manner*' (Sokal and Sneath, 1963, 53).

Now, following this brief description of the aims and methods of phenetic taxonomy, let us look in a little more detail at some of the major stages. First, there is the question of the type of organism to be classified. Secondly, there is the problem of the type of character to be chosen. Thirdly, there is the problem of deciding how to estimate resemblances (i.e. similarities and differences) between complete organisms (i.e. OTUs), given the resemblances between their characters. Fourthly, there is the problem of the type of cluster analysis to be used. Finally, there is the question of how to present one's findings to the world. At this point I shall ignore the first question, and take the pheneticists' subject-organisms as given. In the next section, I shall consider how the pheneticist might choose the sample he classifies. Here, I want to take in turn the succeeding stages starting with the choice of characters to be coded.

Sokal and Sneath call the basic units of information on which they base their taxonomy 'unit characters'. These they define as taxonomic characters 'of two or more states, which within the study at hand cannot be subdivided logically except for subdivision brought about by changes in the method of coding' (Sokal and Sneath, 1963, 65). It is important to note, right at the start, that not every character of an organism can be a unit character. Some are inadmissible. Examples of such inadmissible characters are *meaningless* characters, these being attributes which are not a reflection of the organisms' genotypes (e.g. the names of specimens) and 'characters whose response to the environment is so variable that it is not possible to decide what is environmentally and what is genetically determined' (Sokal and Sneath, 1963, 66), *invariant* characters, these being characters which do not vary within the entire sample, and characters which are *highly correlated empirically*. An example of characters highly correlated empirically is the pink eyes and white skin of albinos. 'The close correlation between pink eyes and white skin of total albinos in most vertebrates would be counted as a single character, since the total absence of pigment implies lack of retinal pigment' (Sokal and Sneath, 1963, 68).

Now, although it would seem that any character other than those definitely labelled 'inadmissible' can be used as part of the foundation of a phenetic analysis, and although, as we saw in the last chapter, the practice of weighing certain characters is considered quite reprehensible, it should not be thought that one can now go straight ahead using the admissible characters as the basis of a resemblance matrix. Certain kinds of character must be treated with care, otherwise one might end with a somewhat distorted estimation of the character's importance. This applies particularly to such a thing as an organism's size (or an organism's part's size), since this is something which can vary quite drastically with age. In cases like this, that is, where one is presented with characters which vary greatly (by size) within a sample, but where one would not want to rule the characters inadmissible on the grounds that their variation was just a function of the environment, it might sometimes be best to use some kind of logarithmic scale of measurement to avoid undue distortion (since, sometimes size is known to vary according to a logarithmic function). Thus, in such an instance, one would not give undue weight to large size just because it was very large. Similarly, care must be taken when differences between organisms might be expressible in terms of one simple mathematical transformation function. For example, in Figure 8.1 one's first reaction might be that the two leaves differ in many ways. However, the transformation grid shows that the shape of (b) is a simple function of the shape of (a). The differences between these two leaves should be expressed in terms of this function, rather than in terms of all the absolute differences that one can find. Hence, it can be seen from these examples that, although weighting as such is not permitted, there are times when a certain 'transforming' of the figures is called for.

Finally, in this discussion of the nature of unit characters, let us note that they should be drawn without bias from as wide a range of aspects of the organism as possible. This range includes morphological characters, physiological characters, behavioural characters, and ecological and distributional characters. Also, as noted earlier, one

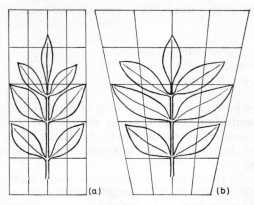

Figure 8.1

(From Sokal and Sneath, *Principles of Numerical Taxonomy*, Freeman and Company.)

should aim for a minimum of about 50 characters. Sokal and Sneath admit that the more characters one studies, the more information one gets; but they commit themselves to what they call 'the hypothesis of the matches asymptote'. After a certain time, the value of the new information gained from each new character falls away drastically, and the search for new characters hardly remains worthwhile.

This now brings us to the next major stage of phenetic taxonomy, namely the problem of estimating phenetic resemblance between OTUs. It would hardly be appropriate here to attempt anything like a thorough analysis of all the various facets involved in the mathematical estimation of phenetic resemblance; however, it should be noted that there do seem to be various approaches, and that whilst none of these is self-evidently the right one, none of these is self-evidently the wrong one. Some approaches have strengths not possessed by others, and some have weaknesses. Many of the approaches rely on a notion of taxonomic distances between the co-ordinates of two points (i.e. as the square root of the sum of the squares of the individual differences). Other approaches, perhaps the most obvious ways of trying to calculate taxonomic resemblances between OTUs, involve what Sokal and Sneath call 'coefficients of association'. Here one tries to express the relationship between two OTUs as a simple fraction, in which one compares some function of the differences and similarities between the unit characters with some other function of the differences and similarities. Thus, for example, suppose one has two OTUs analysed into n unit characters. They agree on m of these characters, and disagree on u. Three possible ways of quantifying the OTUs resemblance-relationship are as follows:

(1) $S_{SM} = m/(m + u) = m/n$

(2) $S_{RT} = m/(m + 2u) = m/(n + u)$

(3) $S_H = (m - u)/n$

The first way just compares the number of matching characters with the total number of characters—the second and third ways give rather more weight to the unmatching characters. (S_{SM}, S_{RT}, S_H are the names of the three different quantifications of the resemblance-relationship.)

As noted just above, phenetic taxonomists would not consider any particular one of these ways of quantifying the relationship between the two OTUs as being *the one unique* way of quantifying the relationship. Rather, they would want to choose the way which fitted best in the particular circumstances. Thus suppose, for example, one had a set of OTUs and one were considering 100 unit characters. If the OTUs were such that the number of matched characters between the OTUs varied right over the scale (i.e. from almost total matching to almost total non-matching), then the simple matching coefficient S_{SM} might be appropriate. One would get ratios from almost 0 to almost 1. On the other hand, if the number of matched characters were bunched up, say varying from 70 up to 100, one might prefer one of the other coefficients, so that the spread of similarity ratios would be more evenly distributed. Presumably, if the number of matched characters were highest in the 90s, and tailed away down to 70, one would prefer the S_{RT} coefficient; but if the number of matches were evenly distributed between 70 and 100, one would prefer the S_H coefficient. The one definite conclusion that one can draw is that, whatever the coefficient chosen, whether it is one of those above or some other, or whether indeed one chooses to estimate resemblance in some way other than by using a coefficient of association, it is clear that phenetic taxonomy leaves the classifier with considerable room for personal choice.

This now brings me to the fourth stage of phenetic taxonomy—the clustering together of the coefficients of similarity. Here again one finds that the classifier is left with considerable freedom, since after considering various ways of clustering Sokal and Sneath conclude that 'each of the clustering methods described in the previous section is valid in its own right if consistently applied' (Sokal and Sneath, 1963, 190). Probably the simplest way of clustering is what they call 'single linkage' clustering. First, this method clusters together those with the highest possible similarity coefficient, then by equal steps it lowers the level of admission. Thus, given five OTUs one might get the following result:

S	OTUs	
0·99	1, 2	
0·98	1, 2, 3	4, 5
0·97	1, 2, 3	4, 5
0·80	1, 2, 3, 4, 5	

Here we have OTUs 1 and 2 joining at a similarity value of 0·99, OTU 3 joining them at 0·98 and OTUs 4 and 5 joining each other at 0·98, and finally at 0·80 we get all five OTUs clustering together.

Now, given this kind of clustering, it should be noted that the admission of an OTU or a cluster into another cluster is by what we may call the criterion of single linkage. By this we mean that if a similarity level of 0·88 would admit an OTU into a cluster, a single linkage at that level with any member of that cluster would suffice to warrant admission. Similarly, any pair of OTUs (one in each of two clusters) related at the critical level will make their clusters join. Thus, while two clusters may be linked by this technique on the basis of a single bond, many of the members of the two clusters may be quite far removed from each other. (Sokal and Sneath, 1963, 180–1)

To remedy this problem, Sokal and Sneath offer more sophisticated ways of clustering. Some ways are variants on the single linkage methods; other ways use more complex methods. One method, for example, requires that an OTU be related with every member of a group before it is allowed to join. (Not surprisingly, this is known as 'complete linkage' clustering.) We need not discuss these other methods in detail, but can merely note that, as usual, considerable freedom is given to the classifier, and as noted above, Sokal and Sneath explicitly state that, in itself, any method is as good as any other.

Finally, there is the problem of presenting these results of clustering. This is frequently done by means of a *dendrogram*. A dendrogram occupies the positive quadrant of a graph—the abscissa having no meaning, being used only to separate the OTUs, and the ordinate being some similarity coefficient scale from 0 to 1 (or 0 to 100 or 1000, to avoid decimals). When 2 OTUs or 2 groups of OTUs have a similarity coefficient which requires that they be clustered together, this is shown by joining together lines stemming down from the OTUs or groups. Although dendrograms are usually shown with the ordinate vertical, there is no reason why it should not be horizontal.

Figure 8.2 is a very simple example of a dendrogram, drawn showing the relationship between seven OTUs.

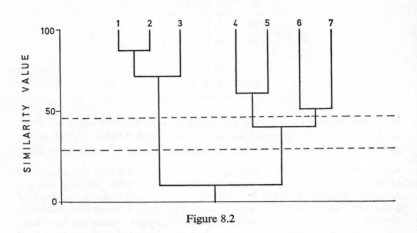

Figure 8.2

As can be seen, OTUs 1 and 2 are linked at about a similarity value of 90, 3 joins them at about 80, and the group consisting of these three finally joins the other grouped OTUs at about 10.

Lines can be drawn on the dendrogram parallel to the abscissa. These are called 'phenon lines' (in Figure 8.2 we have phenon lines at about 25 and 45). Groups which are affiliated at a level not lower than x are said to belong to an 'x-phenon'. Thus, OTUs 1, 2, and 3 (taken together) belong to a 45-phenon, to a 25-phenon, and in fact to a phenon at every level lower than 80. OTUs 4, 5, 6 and 7 (taken together) on the other hand belong only to phena at levels lower than about 40. From this we can see that, in one respect, phenetic taxonomy (inasmuch as it uses dendrograms) is very similar to evolutionary taxonomy. It assigns organisms to non-overlapping classes which are grouped hierarchically, and hence it does not break entirely with the formal structure of evolutionary taxonomy. However, in another respect, phenetic taxonomy does differ greatly from evolutionary taxonomy. Whereas evolutionary taxonomy has only a limited number of levels of taxa into which to put organisms, phenetic taxonomy allows organisms to be put into an unlimited number of phena—in fact just as many as one wants. And, of course, it goes almost without saying that there is a fundamental difference between an evolutionary 'tree' (as in Figure 7.2) and a superficially visually similar dendrogram. The former is intended to show phylogenetic relationships, whereas the latter does not do this at all (although possibly the information it incorporates might later be used as the basis of inferences about such relationships).

This ends my brief exposition of phenetic taxonomy. Let us now turn to the evolutionary taxonomists' criticisms of this new kind of taxonomy.

8.2 The evolutionists' reply to phenetic taxonomy

As might be imagined, evolutionary taxonomists have been no less harsh in their estimation of the worth of phenetic taxonomy than were phenetic taxonomists in their estimation of the work of evolutionary taxonomists. One reviewer of *Principles of Numerical Taxonomy* concluded that, 'After reading this book carefully it is my considered opinion that numerical taxonomy is an excursion into futility' (Ross, 1964, 108). In this section, I intend to consider three criticisms which have been made of phenetic taxonomy. I shall suggest that two can be answered successfully by the phenetic taxonomist, but that the third presents him with far more difficulty.

The first criticism is that phenetic taxonomy is 'meaningless'. This is no doubt a reaction that many evolutionary taxonomists have had and is well expressed by Delevoryas. He writes as follows:

It seems to me that the ultimate goal toward which all biologists should be striving is an understanding of the course of evolution. No matter what new developments are brought to light in molecular biological studies, in ecological investigations, and in countless other biological disciplines, these new ideas are really only a means toward our understanding of a problem on a much grander scale.

In attempts to classify organisms, then, a system based on concepts of relationships and descent would seem to be the most worthwhile goal. If a system is not phylogenetic, it is really meaningless. If a system is not phylogenetic, it really cannot be judged 'better' or 'worse' than another that is not. It is really pointless to argue about the relative merits of various artificial schemes. (Delevoryas, 1964)

Other evolutionary taxonomists have echoed Delevoryas' words. Just one is Simpson, who, when he justifies his allegiance to evolutionary taxonomic methods and principles, declares himself in favour of a 'natural' system, and then concludes that 'in short, if such a thing as natural classification can meaningfully be achieved, it must be by evolutionary classification' (Simpson, 1961, 57).

There are times when it is not easy to escape the impression that when these critics of phenetic taxonomy damn it in this way, all they are doing is labelling, *a priori*, evolutionary classification as the only true 'meaningful' classification, and then condemning phenetic taxonomy because it (not surprisingly) does not live up to their

F

billing. However, assuming that their complaint has a little more force than this, let us try to uncover the real reason behind it. The reason lies, I think, in a difference of opinion about the true purpose of organic classification. Both evolutionary and phenetic taxonomists agree over certain purposes of classification; but they differ over another crucial one. Both sides, for example, agree that a classification provides a *summarizing* device enabling one to file organic collections in some kind of order, so that one does not have to spend hours looking for a desired specimen. Both sides also agree that a classificatory system should have a strong *predictive* value. As soon as, for example, one identifies something as a 'bird', one can make a large number of forecasts about its circulation, skeleton, reproductive habits, and so on. Because one has a classification, one does not have to discover everything anew whenever one finds a new organism—the classification enables one to predict properties of the new organism.

However, although both evolutionary and phenetic taxonomists agree on the summarizing and predictive purposes of a classification, they differ over a third purpose, namely the *explanatory* value a classification might have. For evolutionary taxonomists, a classification must have some kind of explanation built in—something which in some way takes note of the reason *why* organisms are as they are. On the other hand, phenetic taxonomists, whilst not denying that their classifications may be very useful in the search for the reason why organisms are as they are, refuse to accept the stipulation that a classification must necessarily try to reflect the underlying causes of organic diversity. Thus, for example, on the one side we have Mayr arguing as follows:

For the scientist-taxonomist the most important meaning of a classification is that it is a scientific theory, with all the qualities of a scientific theory. First of all, it has an explanatory value, elucidating the reasons for the joint attributes of taxa, for the gaps separating taxa, and for the hierarchy of categories. . . . It is sometimes argued that the descriptive and the explanatory aspects of classifications should be neatly separated. This is impossible. A good classification of organisms is automatically explanatory. (Mayr, 1969, 79)

In opposition to this view, in a pro-phenetic paper by the taxonomist Michener, one section, entitled 'Should the system of classification contain an explanatory element?', concludes that 'In short, then, to answer the question posed in the title to this section, the classification and its explanation should be separate' (Michener, 1963, 79)

It is because of this difference over the purpose of an organic classification that (I would suggest) the charge of 'meaninglessness' arises. No one denies that the reason for organic differences is a genetic one which in some way reflects organisms' pasts; but taxonomists differ over whether or not their classifications should pay attention to this explanatory reason. Evolutionary taxonomists feel very strongly that one should, and hence they feel that any classification which does not is in some important sense 'meaningless'. It does not even try to do one of the most important things a classification should do.

Are evolutionary taxonomists right in condemning phenetic taxonomists for not even trying to incorporate into their systems an explanatory element? My own feeling is that they are not, for, unlike Mayr, I do not see that in itself a system of classification is a theory (although, like evolutionary classification, it might be inspired by and useful to a theory). A classification is a division based on a set of rules and, for this reason, is neither true nor false (which is what a theory is). This is not to deny that if, for example, evolutionary taxonomists can show that phenetic taxonomy is inferior to evolutionary taxonomy in its ability to enable taxonomists to summarize material or to predict things, then in this respect phenetic taxonomy is fair game. The proof of the pudding is in the eating, and if phenetic taxonomists cannot deliver what they claim to be able to deliver, then they are rightly open to criticism. However, it seems a little unfair to condemn them out of hand for not doing something they do not set out to do. One may feel that their systems are limited, and if they fail to be of much value either with respect to summarization or prediction, one may feel their systems are useless; but this hardly makes the systems meaningless.

Moreover, before the evolutionary taxonomist concludes that this is all a squabble over words—'limited', 'useless', 'meaningless', what does it matter, the main point is that a system which is used for prediction without explanation is bound to be no good—he should take note of the fact that some things which have been used for prediction without explanation certainly have been a great deal of good. The Babylonians had incredible success in predicting the motions of the heavens; but they did this entirely in an arithmetical way—treating the celestial motions as discoverable independent variables and computing their various interactions. They did not attempt in any way to explain what they were predicting (unlike the Ionians, who 'explained' the heavens with great gusto, but who could hardly predict a thing). Thus, it would seem that past precedent also dictates that the phenetic taxonomist may legitimately refrain from trying to build explanatory systems. Otherwise, one must refuse to admit that Babylonian astronomy had any worth, and to do this would seem to be a classic example of cutting off one's scientific nose to spite one's philosophical face (see Toulmin, 1961).

The second criticism of phenetic taxonomy that I want to discuss has been made by Ghiselin, who feels that there is something philosophically objectionable about the whole attempt by phenetic taxonomists to quantify the phenetic differences and similarities between organisms. Ghiselin feels that at the heart of this attempt is the mistaken assumption that similarity is some absolute intrinsic property of organisms (just as mass is an intrinsic property of objects). Because of this assumption, phenetic taxonomists think they can assign units to similarity and thus compare organisms' similarity, just as one can assign units to mass and thus compare objects' mass. Ghiselin argues against this position as follows:

Similarity is a relation; things are not twice as 'similar to' any more than they are twice as 'around'. . . . When I say that rats are similar to mice, I do not mean that they are composed to an equivalent number of comparable entities. If someone says that two organisms differ in 75 per cent of a sample of characters, while two others differ in only 65 per cent of these, he cannot meaningfully assert that one of these pairs possess a greater amount of difference, in the same sense that one animal may be said to have more mass than another. One would obtain just as meaningful a figure by adding two oranges, a glass of water, and a telephone number. If one is going to measure, one's units must be equivalent. Organisms are not composed of a finite number of equivalent building blocks which may be considered units of similarity; a long tail and a red nose simply cannot be compared. Therefore, one may erect systems of classification encompassing classes differentiated in terms of similarity in particular characteristics. But no two systems can be compared, except insofar as the parts show corresponding order. For this reason . . . quantitative similarity is a metaphysical delusion. (Ghiselin, 1966, 214)

Now let us note immediately that, apart from anything else, if Ghiselin's criticism is well taken, then Mayr's version of evolutionary taxonomy is in as much trouble as any phenetic taxonomy, since if it is illegitimate to consider phenetic similarity in this way, then it would be no more proper to treat genetic similarity in this way. Although Mayr does not go to the extent of actually putting figures to genetic similarity, he wants to talk of one organism being genetically more similar to a second organism than it is to a third. Moreover if long tailedness and red noseness are genetically caused, he would want to be able to compare these genetic causes in some way which would be very similar to the comparison of the long tail and red nose themselves.

Obviously there is something a little suspect about Ghiselin's argument and it is not too hard to see what it is. Although similarity is not an absolute property of organisms like mass is of objects, the phenetic taxonomist's treatment of phenetic similarity is not dependent on such a supposition. This is clearly shown by the fact that Sokal and Sneath exclude invariant characters from their list of permissible unit characters. They are not trying to measure some absolute property of organisms, but are rather trying to assess organisms with respect to other organisms. Thus, invariant characters which do not vary throughout the whole sample being classified, do not interest them. Now, since phenetic taxonomists are not dealing with some property like mass, their task seems much more hopeful. For them, phenetic similarity is a relationship between organisms, and one can certainly quantify at least some relationships between organisms. Consider, for example, the distance between organisms. This is not a property of organisms like mass— organism *A* is not 'five miles distant'—on the other hand, one can talk of one organism being closer to a second than a third is. Also, one can quantify distance. It makes sense to say that two organisms are five feet apart. *A priori* there seems no reason why phenetic similarity should not be treated in the same way—we certainly do talk in a casual way of organisms in a certain group being more similar to some organisms than to others. Moreover, the obvious way in which to try to quantify phenetic similarity seems to be to try to divide organisms into many small parts and then to compare these. Admittedly, the attempt to do this may run into some difficulties (some of which I shall consider shortly). On the other hand, the one difficulty that Ghiselin highlights, namely that of comparing two different characters (like a long tail and a red nose), does not in itself seem to be a reason for an immediate condemnation of the pheneticists' programme. Things are never exactly the same, and yet we compare them. *A* is farther from *B* than it is from *C*, if *A* is 5 miles from *B* and only 4 miles from *C*. This holds, even if the miles from *A* to *B* are over stony ground and the miles from *A* to *C* over green fields.

Of course, none of this is to deny that the attempt to quantify phenetic similarity may look strange—particularly at first. However, this was probably the case with any first attempt to quantify. No doubt the thermometer seemed very odd in its day. The important point is that it does not seem that Ghiselin can properly rule out phenetic taxonomy even before it starts. At least, he cannot do so solely on the grounds that the notion of quantitative similarity is philosophically unsound. As we shall see, developing the notion raises big problems; but it does seem to be a basis for a viable scientific programme.

I now come to the major criticism of evolutionary taxonomy—the criticism that phenetic taxonomists purge taxonomy of few of the ills which they find in the work of the evolutionary taxonomists. Here,

I think, we have something the force of which cannot be denied. As we saw, phenetic taxonomists accuse evolutionary taxonomists of being insufficiently 'operational', of supplying results which cannot be duplicated by independent investigators, and, above all, of failing to pay proper attention to the canon of 'objectivity'. What the phenetic taxonomists claim is that they will sweep away the evolutionists' subjective morass, and replace it with a system free from the taint of individual biases and wishes. For example, Sokal and Sneath write that phenetic taxonomy (which they call numerical taxonomy)

would have no claim to the serious attention of biologists unless it could overcome some of the faults found in conventional taxonomic procedure. While we feel that the methods to be discussed below have a number of ancillary advantages . . . the outstanding aims of numerical taxonomy are *repeatability* and *objectivity*. (Sokal and Sneath, 1963, 49)

Regretfully, it does not take much ability to see that phenetic taxonomy falls short of these stated aims. To see this in more detail, let us briefly refollow the path of the phenetic taxonomy through its major stages, beginning this time, however, with the choosing of the OTUs to be classified.

Usually, the taxonomist is faced with numbers of organisms which run into the thousands, if not into the millions. Consequently, whatever might be the case in theory, the phenetic taxonomist is going to be faced with the task of choosing certain examples of types to represent their fellows. It is on this relatively small group of examples alone that he will perform his studies. Even with the use of computers he can do nothing else, for no one could (or would) attempt to classify millions of organisms in one study. But faced with the need to select representative specimens, how is the phenetic taxonomist to satisfy it? The only reasonable way to do this would seem to be by falling back on orthodox means. In particular, if one is dealing with organisms from more than one species, one must recognize the legitimacy of the biological species concept and select specimen organisms which represent each and every one of these species. Thus, for example, Michener writes:

If one did not recognize the species at the outset of a numerical taxonomy study one would be faced with the enormous task of a numerical taxonomy of masses of unselected individuals (including thousands of each common kind) to have a chance of including any of the rare kinds in order to discover the species. Gathering data (many characters per specimen) for such an undertaking would be hopelessly wasteful of time. . . . The time-consuming gathering of data for numerical studies should therefore await recognition of the kinds of organisms and development of preliminatory classifications. (Michener, 1963, 159)

In a similar manner, Sokal and Sneath recognize the need to rely on the biological species concept, since they write that 'except for special studies aiming at intraspecific classification, the most customary unit in zoology and botany will be the species' (Sokal and Sneath, 1963, 120).

Now clearly, although Sokal and Sneath later justify their use of the species (by arguing that it is more reliable than other categories), insofar as phenetic taxonomists must rely on the species, their taxonomic systems will suffer from all of the deficiencies that evolutionary taxonomic systems suffer from due to the species. In particular any 'subjectivity' caused by one's species concept will be reflected in their systems. For example, phenetic taxonomists themselves will have to make decisions about where in time to end one species and begin another. This may not be a crippling problem; but it does force the phenetic taxonomist to make decisions which he claimed his system would make unnecessary.

Next in phenetic taxonomy comes the choosing and evaluating of unit characters. This clearly involves blatant deviation from the norms of repeatability and objectivity (in the way that they seem to be conceived of by Sokal and Sneath). The whole point of phenetic taxonomy is that through it one can 'approach the goal where different scientists working independently will obtain accurate and identical estimates of the resemblance between two forms of organisms, given the same characters on which to base their judgement' (Sokal and Sneath, 1963, 49). All kinds of personal bias, or wish, or interest, are to be removed. The individual taxonomist is to give up his own free choice, and for this he will be presented with the unadorned truth. Moreover, in order to carry this programme through, attention is to be paid exclusively to the physical characteristics of organisms. All theorizing, such as a taxonomist like Mayr performs, is to be strictly eschewed. Any speculation about the past or genetic backgrounds is to come after the classification—not before or during it.

In the light of these ideals, it seems quite incomprehensible that Sokal and Sneath should state them, and then turn around and give all kinds of recipes for omitting or adjusting the relative values of different unit characters. Consider, for example, the directive to ignore variable characters, such as the number of leaves on a tree. Suppose some taxonomist wants tree-leaf numbers to be a unit character. There just seems to be no way in which one can stop him, *short of appealing to genetic considerations*. The number of leaves a tree has is just as much a phenetic property of the tree as the shape of the leaves. But if one appeals to genetic considerations (which incidentally is just what Sokal and Sneath do), and if one tries to show that exact tree-leaf number is not a function of the genes in quite the way that (say) leaf-shape is, then one is going beyond a phenetic classification. This may be a legitimate way of classifying —Mayr claims it is—but it is no longer phenetic classification. On the other hand, if one is not prepared to appeal to genetics, then one has no right to stop someone using exact tree-leaf number as a unit character.

This conclusion seems to apply even more strongly to the problem of empirical correlations. If one does not appeal to the genes (which Sokal and Sneath do), then in the case of total albinos one has no way of deciding whether pink eyes and white skin together count as a unit character, or whether they should count apart as unit characters (or whether the two eyes should count apart, or whether each white hair is a unit character, or so on and so on). Here again, phenetic taxonomists are caught in a dilemma. Go beyond the phenetic evidence and appeal to the genes, in which case the difference between a phenetic taxonomist and an evolutionary taxonomist like Mayr disappears, or refuse to go beyond the phenetic evidence, in which case the chances of repeatability and objectivity sink drastic-ally. There just is no way on the surface of deciding in a completely definitive manner whether to treat white skin and pink eyes as one character or two. Moreover, exactly similar considerations apply to Sokal and Sneath's other directives about unit characters. If one does not lay down rules in the way that they do, then hopes of complete repeatability and objectivity are lost. On the other hand, if one does lay down such rules, then it is difficult to see how one's classification can avoid going beyond the purely phenetic to the genetic side of organisms. Either way, the goal of phenetic taxonomy is put beyond one's reach.

The next stage of phenetic taxonomy involves the estimation of resemblances between OTUs on the basis of the difference in their characters. As we saw, the more inventive the phenetic taxonomist, the more ways there are open to him to estimate these resemblances, and there seems to be nothing absolutely forcing two taxonomists to take the same way. Sokal and Sneath themselves admit this when they concede the possibility of viable alternatives. However, if the phenetic taxonomist does have alternatives to choose from, then it would hardly seem that his subject achieves the repeatability and complete freedom from personal decision that we are told is the hall-mark of phenetic taxonomy.[1]

[1] One important point should be noted here. The reader may think that I am denying the possibility of finding a quantitative measure of phenetic resemblance. If so, it is somewhat strange that I should earlier have criticized Ghiselin for denying the possibility of finding such a measure. To put the record straight—I

An almost identical argument applies to the final major stage of phenetic taxonomy, where the OTUs are clustered together. The phenetic taxonomist has many choices, and the decision between, say, adopting a single linkage method of clustering rather than adopting a complete linkage method seems no different in principle from the evolutionist's decision to group contemporaneous groups rather than ancestor–descendent groups. Both decisions involve the desires of the classifier—desires which may or may not be shared by another classifier. Hence, at this concluding stage, as before, the pheneticist faces choices no less than does the evolutionary taxonomist.

Of course, phenetic taxonomists themselves are not unaware of the problems faced by their taxonomy. Even Sokal and Sneath, despite their many claims about repeatability and objectivity, finally qualify their position by saying that they 'do not claim that numerical taxonomies are objective realities; the fact that a number of slightly differing taxonomies may be obtained by different statistical methods is clear evidence that they are not' (Sokal and Sneath, 1963, 268). However, merely acknowledging the problems hardly makes them go away.

It is clear by now that just as difficulties and choices occur in evolutionary taxonomy, so they occur in phenetic taxonomy. Hence, both taxonomies leave open degrees of freedom to the individual taxonomist. Nevertheless, as we leave the criticisms of phenetic taxonomy and turn to the final section, it is important to note one thing. The choices faced by phenetic taxonomists have their counterpart in choices faced by evolutionary taxonomists. Consequently, whilst the pheneticist should not condemn the evolutionist for his problems arising from the need to make decisions, so neither should the evolutionist condemn the pheneticist for his problems arising from the need to make decisions. Organic classification is a practice which leaves freedom for the individual practitioner, and this is a truth which holds, whichever way one tries to classify. Moreover, whilst it is obviously the case that one should try to eliminate sources of bias due to the preconceptions of the taxonomist,

---

am not denying the possibility of finding a quantitative measure of phenetic resemblance, nor would I deny that some measures accord more with common intuitions than do others. Indeed, I am quite sure that some measures do accord more with common intuitions than do others. But, I confess that I would deny the possibility of finding a *totally objective* measure of phenetic resemblance, in the sense that this measure is given to scientists rather as God gave the Ten Commandments to Moses. However, for reasons which I gave in Chapter 7, I do not find this failure to find such an objective measure very worrying. This is because I do not think any science is as 'objective' as Sokal and Sneath would have taxonomy.

one should not sacrifice everything on the high altar of 'objectivity'. Neither within taxonomy nor outside of it, does or could the totally objective science exist.

8.3   *Is there a proper way to classify?*

Rival taxonomists speak of each other's work in a way unknown outside of the House of Commons. ('Nominalist' and 'Platonist' are favourite epithets and are hurled by both sides with parliamentary vigour.) Nevertheless, although there are obviously very important differences in the two taxonomies, as one tries to summarize the findings of the discussion in this chapter and the last about evolutionary and phenetic taxonomy, it seems proper first to point to the strong similarities between the taxonomies. For a start, both sides share the same overall aims for taxonomy: they both want taxonomy to be a scientific discipline which can stand comparison by the standards of other sciences. Mayr, as we saw, claimed that 'for the scientist–taxonomist the most important meaning of classification is that it is scientific theory, with all the qualities of a scientific theory' (Mayr, 1969, 79). Similarly, a major theme of the phenetic taxonomists is the need to eliminate the subjective bias of individual taxonomists, and to make taxonomy 'objective' like other sciences. Of course, this is not to deny that the ways in which the two sides try to make their taxonomies like other sciences are quite different. The evolutionary taxonomists think that a science must have explanatory value. The phenetic taxonomists deny this; but, on the other hand, they in turn espouse supposed 'scientific' ideals downplayed by evolutionary taxonomists. In particular, the phenetic taxonomists seek repeatability, total objectivity, and conformity to the principles of operationalism.

A second point of resemblance between the two taxonomies is that both sides seem ambivalent about their more immediate aims. As we saw, evolutionary taxonomists differ over what they seek. Some, the geneticists like Mayr, want classifications to reflect the underlying genetic nature of organisms. Hence, strictly speaking, phylogeny is irrelevant for them—although, not entirely consistently, Mayr does require that his taxa be monophyletic (that is, he wants the members of no taxon of rank n to be descended from the members of two or more contemporaneous taxa of rank n or higher). Others, the genealogists like Simpson, want their classifications to pay fairly direct attention to the phylogenetic history of organisms. Hence, for them, classifications must not be 'inconsistent' with phylogenies. Although the aims of the geneticist and the genealogist will frequently coincide, in certain cases they may not. On top of this confusion, it must be further recognized that, supposing one adopts a position like Mayr's, there is still work to be done to clarify one's aims. Despite granting that the notion of 'genetic similarity' does not seem as untenable as some (like Ehrlich) would have it, additional guidelines need to be laid down before it can be a truly satisfactory basis for organic classification. On the other hand, although the phenetic taxonomists rightly point to this kind of problem with evolutionary taxonomy, they too need to put their own house into order. Probably the main aim of phenetic taxonomy as it is presently conceived is to produce a taxonomy which reflects overall phenetic resemblance. For example, Sokal and Sneath write that in their taxonomy they aim for 'the numerical evaluation of the affinity or similarity between taxonomic units and the ordering of these units into taxa on the basis of their affinities' (Sokal and Sneath, 1963, 48) and later they assure us that by the term 'affinity' they 'imply a solely phenetic relationship' (Sokal and Sneath, 1963, 123). Similarly, Michener claims that one of the objectives of classification 'should be to reflect as closely as possible the resemblances among kinds of organisms' (Michener, 1963, 153). However, phenetic taxonomists are far from entirely consistent in their writings about the major purpose of their taxonomies. Sometimes, indeed, one gets the feeling that there is no unique, major purpose, and that any method of classification can be used, so long as it meets the particular needs of the moment. Thus, 'having defined precisely what is pertinent to biological classification and what are the best statistics for achieving a given end, we can obtain taxonomies which fulfil the needs for which they are devised' (Sokal and Sneath, 1963, 268–9). It is obvious that, at this point the phenetic taxonomists, no less than the evolutionary taxonomists, need to do some careful rethinking about what they consider to be the immediate aims of their taxonomy. (Of course, it may be that phenetic taxonomists really do not want any specific, unique purpose guiding their taxonomic work. In itself, such a pragmatic stand does not seem to me to be objectionable; but I do think that if this is to be their position, it needs spelling out in an explicit manner.)

A third point of resemblance between the two taxonomies is the latitude and responsibility that they give to the individual classifier. Evolutionary taxonomists openly admit this fact—Simpson, for example, speaks of classification being as much as 'art' as a science —but even the phenetic taxonomists, despite their usual claims about repeatability and objectivity, leave many decisions in the hands of the classifier. Since I have already had so much to say about this, I shall say no more here.

As I said, in drawing attention to the similarities between the two taxonomies, I do not want to minimize their differences. The decision to base classification on phylogeny or genetics rather than phenetics (or vice versa) leads to differences in both the practice and results of classification. This being so, let me now ask my final question about taxonomy. On the basis of the total discussion in these two chapters, can we say which taxonomy biologists ought to adopt? Unfortunately, short of putting the two taxonomies to fairly extensive practical tests and seeing if either or both do in fact prove to be useful summarizing and predictive devices (as they both claim to be), it is not easy to say very much. However, strictly on the basis of the past discussion, one starts to suspect that the answer to the question of which is the better taxonomy is that 'it all depends'. We have seen that both taxonomies have problems which have to be solved or avoided as they turn up. On the other hand, we have also seen that neither taxonomy is really quite as bad as its opponents would have it. This being so, I think the better taxonomy probably depends on one's area of study. If one is studying organisms with a good fossil record, for example the mammals, then perhaps evolutionary taxonomy is the more obvious choice. The inductive generalizations one will have to make to infer unknown phylogenies or genetic resemblances will have a relatively large amount of supporting evidence. It is, perhaps, no surprise that biologists like Simpson who are the strong supporters of evolutionary taxonomy are just those taxonomists whose interest does lie in groups like the mammals. It is no surprise also that the differences between evolutionary taxonomists is reflected in their different biological interests. Simpson, who emphasizes phylogeny in classification, is a paleontologist. Mayr, who emphasizes genetics, is a neontologist.

Alternatively, if one's interest lies in organisms with little or no fossil record, or in organisms where evolutionary taxonomic principles are really difficult to apply (e.g. asexual organisms), then possibly phenetic taxonomy offers one the best way of classifying. As pheneticists perpetually point out, with groups like these one's classification will probably be *de facto* phenetic anyway, whatever one calls it. Again, it is not surprising to find that the supporters of phenetic taxonomy are just those whose interests lie in groups without fossil records. Ehrlich, for example, is interested in butterflies. Sneath is interested in bacteria. The University of Kansas Entomology department is a stronghold of phenetic taxonomy. Of course, phenetic taxonomy is a comparative newcomer—estimation of its overall value will have to wait until it has had a fairly long trial.[2]

[2] In this context, the reader might care to look at the results of a fascinating experiment by Sokal and Rohlf (1970) in which they attempt to classify organisms using people who are quite untutored in taxonomy and ignorant about the ultimate aims of what they are doing—'intelligent ignoramuses',

Perhaps it will turn out that to make good predictions, pheneticists will have to bring their system closer and closer to evolutionary taxonomy (as I think Sokal and Sneath do, when they discuss empirically correlated characteristics). Whatever may be the case, as things stand at the moment neither taxonomy seems perfect; but both seem to be legitimate tools for the classifier of organisms.

9

THE PROBLEM OF TELEOLOGY

Things happen, 'effects', because other things, 'causes', make them happen. A great many explanations in the physical sciences are 'causal' explanations, in that together with laws they consist of the specification of a cause in the *explanans*, the effect of which is cited in the *explanandum*. But this means that in such explanations, the *explanans*-phenomenon cannot come later (in time) than the *explanandum*-phenomenon. This is because in the physical sciences there seems to be a curious asymmetry about causes. If something, *A*, is said to be the 'cause' of something else, *B*, then *A* is temporally prior to or simultaneous with *B*. *A* does not come after *B*. Why should this be so?

One reason might be that the idea of a future cause is logically contradictory, like, for example, the idea of a square circle. However, although some philosophers seem to dismiss future causes for this reason, to me it is far from obvious that the idea of a future cause does contain a straightforward contradiction. Suppose I drop a piece of chalk 10 feet to the ground. Why is it inherently contradictory to suppose that the chalk's motion 5 feet above the ground is caused by the hitting of the floor, or by the chalk's motion just before it hits the floor? Such a future cause does not seem unimaginable in a way that obvious contradictions do. Remember how in *Alice Through the Looking Glass* the White Queen cries only before she has pricked her thumb. Children and philosophers find this passage amusing, pointing to the fact that some (albeit topsy-turvy) sense can be made of future causes. To my knowledge, not even Hegelians find the idea of square circles funny.

Another reason one might give against future causes is that they cannot be used in covering-law type explanations (by virtue of the covering-law requirements of such explanations); but this also seems not to be true. Suppose, to take an example of Hempel (1965), that one wants to explain why a ray of light starting at A in one medium and arriving at B in another medium goes through point C on its way. One might do this by appealing to Fermat's principle of least time (together with other relevant information such as knowledge of the refractive indices of the two media) deducing that the quickest route from A to B is through C. In this explanation, the *explanandum* event (i.e. passage of ray through C) is explained by reference to an *explanans* which contains talk of events before (i.e. ray starting at A) and after (i.e. ray arriving at B) the event of the *explanandum*. We do not think of the arrival of the ray at B as being as a cause of the passage through C; but, from the viewpoint of the covering-law aspect of the explanation, there is no reason why the arrival at B should be less of a cause than the start at A (which latter we do think to be a cause).

The real reason for the physical scientist's distaste with future causes is, I think, the 'problem of the missing or unachieved goal-object'. Suppose, for some reasons, the chalk never hits the ground, the Queen never pricks her thumb, and the ray leaves C but never gets to B. If we allow the possibility of future causes, then we seem to be stuck with saying that the falling chalk, the crying Queen, and the travelling ray are all being caused by things which do not and will never exist. This kind of position does seem to be absurd if not outrightly contradictory. But what alternatives are there open? One thing one might say is that there must be some other future cause, one which will really exist. The trouble with this is that one then seems to lose all control and check over one's scientific hypotheses. Almost anything might occur, and it is hard to see how one could prevent the supporter of future causes claiming that it (i.e. anything) was the cause of a past event. The other thing that one might say is that when the 'goal-object'—the chalk hitting the ground, the pricked thumb, the ray at B—is missing, no future causes are involved. Just the past and present are needed to give a complete causal explanation. The trouble with this is that the doctrine of future causes now seems a bit redundant. One has two identical situations, say of chalk in mid-flight. The one case is caused by the past, the other case is caused by the future; and until the chalk hits or fails to hit ground, one cannot say which is which. But why bother to weigh down one's science with such a cumbersome ontology? Why not just treat both cases as if the causes came out of the past, and forget about imperceptible future causes? It hardly seems that one's science will be any the poorer.

It is for reasons like these, that I think that physical scientists prefer to stick with causes from the past and present. Belief in future causes has no place in physics and chemistry, and hence in causal explanations (i.e. those specifying the cause in the *explanans* and the effect in the *explanandum*) we are explaining by reference to the past or present and not the future. And indeed more generally, in the physical sciences, although one does get cases like the above use of Fermat's Principle, when explanations are given concerning causes and effects, causes and only causes tend to get cited in the *explanans*, and thus all understanding involving causes usually involves explanatory reference to the past and present, and not to the future. (Like nearly every other general statement in this book, this claim must be qualified slightly. At least one physicist recently has toyed with the notion of 'backwards causation' (Csonka, 1969). But Csonka does not wrestle with the above-mentioned philosophical problems with future causes.)

However, when we turn to biology, we find altogether less aversion towards talk which in some sense involves understanding through reference to the future (whether future-causal or not). Biologists are always talking of one thing occurring 'for the sake of' another, or occurring 'for the purpose of' something, or occurring 'in order that' something might happen, or serving the 'function' of something else. In all of these cases we apparently have one thing which in some way at least is directed towards another future thing, which other thing in turn throws explanatory light on the earlier thing. Now, one presumes that the problem of the missing goal-object is no less worrisome for the biologist than it is for the physical scientist. A question of some philosophical interest, therefore, is that of showing that the biologist's explanatory references to the future are not really future-casual or in any other way objectionable. This problem of trying to analyse and understand (i.e. explain) via the future—the problem of 'teleology'—will be my concern in this chapter. (The literature on causation, specifically future causation, is immense. Mackie, 1966, makes some nice points and contains a good bibliography.)

I shall begin by considering one class of phenomena which many think most clearly involve one in problems to do with teleology, and I shall show that, whatever else may be the case, these phenomena do not necessitate an analysis postulating future causes. Secondly, I shall consider what we mean when we talk of another type of teleological (or possibly quasi-teleological) phenomenon, namely that which serves a *function* in a system. Thirdly, I shall consider the nature of biological teleological explanations and the questions of their adequacy. It will be my final claim in this chapter that although biological thought does not presuppose a 'strong' teleology (i.e. a teleology postulating future causes), in a somewhat weaker sense biology has an untranslatable (although perhaps not unremovable) teleological element. In a very real manner, biologists do get explanatory understanding by reference to the future.

9.1   Goal-directed phenomena

The kind of phenomenon in the organic world which has probably attracted most attention from philosophical writers discussing problems stemming from biological teleology is the so-called 'goal-directed' or 'directively organized' phenomenon. In this very wide-spread kind of phenomenon or system we apparently have some event or state, the 'goal', towards which the organism (or group of organisms) is directed (or which, the organism is already in), and where, even though obstacles (literally or metaphorically) stand in the organism's way, the organism will in some sense persist towards this goal (or where the organism will persist in regaining the goal if the obstacles take the organism out of the goal-state). A typical example of such goal-directedness is afforded by the phenomena of sweating and shivering. A human's normal bodily temperature is just over 98°F. If the body gets overheated or overcooled, then things start to happen, namely sweating or shivering, 'in order to' bring the body back to a normal temperature. And here, if anywhere, it would seem that we have future causes at work. The final state (i.e. normal temperature) brings about the sweating and shivering, which, in turn, exist in order to bring the body back to the right temperature.

It is easy to show that, just as in the case of the physical sciences, the problem of the missing goal-object reveals such a (strong) teleological analysis to be inadequate. Sometimes human bodies get heated beyond the stage where sweating can prove effective, and similarly, they get too cold for shivering to help much. Hence, even were one to hold a (strong) teleological position, in the case of, for instance, the martyr being burned at the stake, since (in a successful execution) his living body will never regain its normal temperature, one could hardly invoke the normal temperature as a cause of his sweating as the flames first start to lick around the faggots. Thus, it seems that an alternative analysis of goal-directedness is needed.

From a very general viewpoint, it does not seem that such an analysis is unduly difficult to find; but this is, of course, not to say that in a particular case it might not prove very difficult to give a detailed analysis of goal-directness, nor is it to say that such a particular analysis might not prove very complex. Following Nagel (1961), I would suggest that a system is goal-directed or directively organized if the following conditions obtain. Suppose we have a system which either exhibits some property or mode of behaviour G, or which, if left untouched, would at some later point exhibit G (more briefly, suppose that we have a system which is 'in a G-state'). Suppose that the system can be analysed into a number of different independent parts or processes and that were one of these parts altered on its own and were there no compensating alterations in the other parts, the system would cease to be in a G-state (i.e. G would be lost, and, without compensating move, would never be regained). However, suppose that in the system, whenever at least some of these initial alterations occur (call them 'primary variations') other parts in fact compensate (by what we might call 'adaptive variations') in order to bring the system back into a G-state. A system like this is, I think, properly called 'goal-directed', and it should be noted that it has not been necessary to postulate the existence of future causes, nor has it been necessary to suppose that the goal (i.e. achieving or returning to a G-state) must always be reached. It is quite possible, indeed, as far as the organic world is concerned, invariable, that some primary variations will prove so disruptive that the system has no adaptive variations strong enough to return it to a G-state. The goal could be lost forever.

Before going on, in order to avoid objections a couple of points should be made about this characterization of a goal-directed system. First, it cannot be denied that many such systems may exist in the non-organic world—my concern here is only in the organic world. If there be any real differences between living and non-living things, I do not think they lie here. (Later, I shall mention where I think they do lie.) Secondly, I think that, as in most cases, our idea of goal-directedness is a little blurred at the edges. Suppose a primary variation took a system out of a G-state, and that although this set off response variations (e.g. the martyr's sweating as the fire heats up), these were not strong enough to bring the system back into a G-state. I think possibly we might speak of the system being goal-directed (towards a particular goal) if the response variations were normally adaptive (in the sense that they would bring the system back to a G-state) even though they could not bring the system back to a G-state in this case. If the response variations were not normally adaptive (e.g. the burning of the martyr's flesh as the fire heats up), then the system would not be goal-directed in this respect even in a loose sense.

Now, I admit candidly that the trouble with the discussion so far is that it is all a little bit too abstract. As Taylor (1964), one of the strongest contemporary advocates of some kind of legitimate teleology, asks, how can we be sure that real-life situations can in fact be analysed according to this model? The non-teleologist just seems to have specified the kinds of conditions he would like to find in the world, and then, *a priori*, he claims that such conditions do hold in the world. And even those who in no sense seem to want to adopt a position supporting teleology seem less than comfortable with such an abstract analysis as that which I have just offered. For example, Gruner (1966) feels that all consideration of goal-directedness along the lines just given has 'too much of an anthropomorphic flavour', and I think that he would just like to avoid goal-directedness altogether.

To be honest, I do not really see how one could persuade the determined critic of the analysis I have just offered, other than by a case-by-case consideration of every instance of goal-directedness occurring in biology, and I doubt if we should ever be able to do this in practice, even if we had the time and energy. Some situations are just too complex. However, one can offer some fairly completely analysed instances of goal-directedness showing that they do fit the model, and I think that even a little success makes an objection like Taylor's seem less ominous. Moreover, one can certainly show that Gruner's fears are quite unjustified—such an analysis smacks not at all of undue anthropomorphism. As an instance of a goal-directed phenomenon which can be analysed along the lines suggested, I offer the following.

Many species of birds have a remarkably stable clutch-size. For example, the clutch of a petrel is normally 1, of a pigeon 2, of a gull 3, of a plover 4, and so on. This is obviously not something which occurs by chance, since the egg-number characteristic of a species is far too constant for chance, and in any case, it is clear that most of the birds are actually laying below their physiological limit. If an egg is removed, the bird will bring the number up to the right clutch-size. Furthermore, whatever it is that is keeping the clutches at the sizes that they are must be something essentially genetic. As is well known to poultry breeders, rate of egg-laying is something very directly under the control of the genes—alter the genes in any significant fashion, and this will be reflected in an alteration in egg-number. However, even though the chances of an individual mutation are very slight, overall one would expect some mutations in a particular species, enough to cause some change in clutch-size. Consequently, it would seem that we have here an instance of goal-directedness. The plovers, for example, are in a *G*-state, namely that of having a clutch-size of 4, and despite disruptions (particularly any genetic disruptions) they persist in this *G*-state. Hence, an analysis is required.

The analysis offered by evolutionists is that the particular number of eggs laid by a species like the plover is a result of selection. Birds are best capable of raising a particular number of young—4 in the case of the plover. Any more young, and they will get insufficient food and parental care, and hence not be as fit as birds from a clutch of 4, and this lack of fitness will outweigh the increase in number. Any less young, and there will not be as many as the birds from a clutch of 4 (but, bird for bird, those from the clutch of 4 will not be appreciably less fit than those from smaller clutches). Thus there is strong selective pressure in the direction of a specific clutch-size. (See Lack, 1954, for full details. Wynne-Edwards, 1962, would not accept this analysis, arguing instead for some kind of 'group selection'. But see Lack's effective counter in Lack, 1966, and Williams', 1966, devasting critique of group selection.)

The question which now needs answering is whether or not this analysis offered by the evolutionist fits the model for goal-directed behaviour offered by Nagel. It is fairly obvious that it does. We have systems, namely populations of species of birds. These systems can be analysed into different independent parts, namely the individual birds. At a certain time of year these birds exhibit a specific property, namely the laying of eggs in certain clutch-sizes. At other times of the year, the birds do not exhibit this property, but if left untouched and given time, they will exhibit the property again (i.e. in the next spring). Hence, the systems are in G-states. Suppose now that one of these parts of a system (i.e. a bird) is altered—say one of the birds carries a new mutation, so that it has a different clutch-size from the norm. For example, if it is a plover, then it might have 3 eggs or 5 eggs rather than the normal 4 eggs. This is a primary variation. If nothing else changes, then the system will no longer (and never will again) have the G-state it was in previously. The clutch-size of at least some of the members will always be different from 4 (we are assuming fairly normal inheritance, a large population, and random interbreeding—in fact, of course, to increase the chances that the abnormal genes are passed on it would be best to assume mutations affecting clutch-size in more than one bird). But what happens in reality is that the rest of the system compensates for this change in the clutch-size of one bird, and the compensating changes are enough to bring the system back into the original G-state (e.g. state of laying 4 eggs). In particular, what happens is that certain other parts of the system (i.e. other birds) do things which they would not have done had the original disruption (i.e. the mutation) not occurred. Because of the disruption, these other birds (with normal clutch-size tendencies) will reproduce, when they would not normally have done so. This is because the mutant bird will either have fewer offspring or, if it has more offspring, have weakened offspring. Assuming the first alternative, other birds will be able to reproduce because there will be fewer of the mutant's offspring against which to compete, and assuming the second alternative, the other birds will be able to reproduce because they will be stronger than their competitors (possibly they will be able to push the competitors out, or possibly the competitors will fall by the wayside for other reasons, e.g. the weather). In either case, those birds genetically programmed for normal clutch-size will have a reproductive advantage over the mutant, and because of this, they will be the ones which are selected (or whose offspring are selected). Hence, the population will gradually be brought back to a state of exhibiting G (i.e. normal clutch-size), through adaptive variations.

It is clear that we have here a paradigmatic example of a goal-directed system as defined by Nagel. Even though, on first sight, the future G-state apparently controls the earlier adaptive variations, in fact there is no need to appeal to future causes at all. Nor need one be worried if, once lost, the G-state is never regained (i.e. if we have a missing goal-object). Suppose *all* of the birds of a population mutate from, for example, a clutch-size capacity of 4 to one of 12. In this respect none of the birds will then be any fitter than any of the others, and they could well all over-extend themselves in one season, and thus the population could die out without ever regaining a clutch-size of 4. The theoretical analysis which has been offered can accept this kind of situation. Finally, let us note that the analysis in no sense implies undue anthropomorphism, and thus Gruner's unease at the thought of non-human goal-directed systems is unwarranted. (This is not to deny that in using a word like 'goal' we are using a word first used in the human world. This, in itself, is hardly objectionable. What would be wrong would be if we were transferring things like desires and wants and conscious intentions straight into the non-human organic world. This we have not found it necessary to do, although I myself suspect that some of the higher animals do have a limited human-like consciousness.)

Whilst one could probably write quite a bit more directly about the nature of goal-directed systems, I want now to turn to another aspect of the problems of teleology in biology, namely the problem of what a biologist means when he speaks of something serving or having a 'function'. As we shall see almost at once, our knowledge gleaned in this section about goal-directed systems will not be wasted.

9.2 *Functional statements*

If I say that something has or serves a particular function, then I am saying something which is in some sense (apparently) teleological. For example, if I say that the function of the beating heart is to circulate the blood, then what I am saying is that in some sense the (earlier) beating heart exists for the sake of the (later) circulating blood, or exists for the purpose of circulating the blood. Assuming, for the simple reason that sometimes functions are unfulfilled, that the teleology of a functional statement is not what I have called 'strong' teleology (i.e. that there is no presupposition of future causes in our understanding) the question still remains of how we are to unpack a functional statement, showing that there is nothing objectionable or mysterious about such a statement. The most widely known attempt at such an unpacking is that of Nagel (1961), and so I shall begin by considering this. It will be my contention that Nagel is not merely wrong, but looking entirely in the wrong direction. As soon as his fundamental misconception is revealed, then I think that the correct translation of a functional statement can be found without too much difficulty. (Such a translation will eliminate the talk of functions; but this is not to say that it will eliminate all reference to the future.)

The example of a functional statement around which Nagel conducts his analysis is:

The function of chlorophyll in plants is to enable plants to perform photosynthesis (that is, to form starch from carbon dioxide and water in the presence of sunlight). (Nagel, 1961, 403)

Nagel argues that this statement is equivalent to two statements:

(1) Chlorophyll is necessary for the performance of photosynthesis in plants.
(2) Plants are goal-directed.

I assume, from his examples, that the goal Nagel has in mind is the survival and reproduction of plants.

Now most criticisms of Nagel concentrate solely upon the first translation statement, the argument being that conceivably plants could perform photosynthesis without chlorophyll, that is that chlorophyll is not necessary for photosynthesis (e.g. Lehman, 1965a). However, it is unclear how devastated Nagel would be by this type of criticism, since he does in fact raise the criticism himself, and, whilst admitting that it is in part well-taken argues that 'although living organisms (plants as well as animals) capable of maintaining themselves without processes involving the operation of chlorophyll are both abstractly and physically possible, there appears to be no evidence whatever that in view of the limited capacities green plants possess as a consequence of their actual mode of organization, these organisms can live without chlorophyll' (Nagel, 1961, 404). Perhaps the easiest solution to criticisms of (1) is to admit that the choice of the term 'necessary' is somewhat unfortunate, and to replace the statement by the true

   (1') Plants perform photosynthesis by using chlorophyll.

After all, even though Nagel's reply is probably well-taken and chlorophyll is in some sense necessary, from a more general viewpoint, if we were to say that one thing x served a function y, it is doubtful that we would necessarily want to say that x was necessary for y. Suppose there were another thing, x', which was an alternative way of getting y. We would still say that the function of x was to get or do y (and, similarly, the function of x' was to get or do y), whilst admitting that neither x nor x' alone was necessary for y. On the other hand, we do clearly claim that x, in some sense, gets or does y. One could hardly say that x had a function involving y, if x never had any part in bringing y about.

Clearly the above functional statement (i.e. that chlorophyll's function is bringing about photosynthesis) must imply more than (1'), otherwise all kinds of true statements of the same form as (1') could be translated into false functional statements. Thus consider the statement:

   (3) Long hair on dogs harbours fleas.

Even though (3) is true, no one would want to claim:

   (4) The function of long hair on dogs is to harbour fleas.

Nagel's second statement is supposed to prevent this kind of move, and it is here I suggest that his gravest error lies. The error is not so much that the appeal to goal-directedness fails to rule out every type of incorrect translation, although it certainly fails to do this, since Nagel would have to allow (4), because not only is (3) true, but so also is the statement that dogs are goal-directed towards survival. Rather, the error is that the appeal to goal-directedness is altogether inappropriate. To see why this should be so, let us consider the kind of evidence which would lead us to accept (4) as true.

Suppose someone had kennels in which they kept Afghans (long haired) and Salukis (short haired). Suppose also that they found that on the whole the Afghans lived longer than the Salukis, so much so that, because of this discrepancy in lifespan, the number of offspring per Afghan was significantly higher than the number of offspring per Saluki. Suppose that it was discovered that the reason why Afghans have a greater life-expectancy is because the Salukis contract some kind of mysterious incapacitating parasite carried by a certain species of insect, and that Afghans are not affected by the parasite because they carry more fleas than the Salukis (because of their longer coats) and fleabites provide immunity from the parasite. Finally, suppose that some biologist or veterinary surgeon heard about this, and by further investigation discovered that it seems to hold generally for both domesticated and wild dogs that, in regions where the parasite is common, long hair (because it harbours fleas) tends to increase life-span (and potential reproductive power). I think if he were to report this astounding conclusion to the world, no one who accepted his evidence would deny him if he said 'The function (or at least, one of the functions) of long hair in dogs is to harbour fleas'.

Why should the above hypothesis about the links between long hair, fleas, and immunity from disease make any difference to our feelings about the appropriateness of saying that the function of long hair is to harbour fleas? It seems clear that the reason is that, given the truth of the hypothesis, harbouring fleas has been shown to play an important role in the survival and reproduction of the particular species under discussion. That is to say, we would expect a group of dogs which harboured fleas to have a better chance of survival and reproduction than another group, identical in every respect except that they did not harbour fleas. Of course, since the hypothesis is false, it is easy to see why at present we definitely do not think that a function of long hair in dogs is to harbour fleas. If anything we believe the reverse, since fleabites can be detrimental to the health of a dog.

Returning to a more general level, we can now see that the claim 'The function of x in z is to do y' implies that y is the sort of thing which aids the survival and reproduction of z. Now this is the kind of thing which, as we have seen, biologists call an 'adaptation'.[1] (Actually, up to now, we have usually spoken of adaptations as being actual phenotypic characteristics rather than as the process or products brought about by the characteristics. However, the term 'adaptation' seems extendable in this way, although no crucial aspect of the present analysis rests on such an extension.) Consequently a functional statement in biology draws attention to the fact that what is under consideration is an adaptation or something which confers an 'adaptive advantage' on its possessor. Hence Nagel's example of a functional statement is also asserting,

[1] Wright (1972) in criticizing an earlier version of this analysis of functional statements by me, Ruse (1971c), points out that one must guard against allowing as functions 'one-shot' items or processes which aid survival and reproduction. If a y helped reproduction just once (with no expectation of a repeat) we would probably dismiss it as an 'accident' and would not talk in terms of function. Only if y is preserved and perhaps improved by selection is function-talk appropriate, and, of course, only under such circumstances would we talk of 'adaptation'.

(2') Photosynthesis is an adaptation; that is, plants which perform photosynthesis are more likely to survive and reproduce than plants which are identical with them in every other respect but which do not perform photosynthesis.

(One could, as I have just pointed out, alternatively interpret Nagel's example as pointing to the fact that chlorophyll is an adaptation, although one would also have to indicate that it is an adaptation by virtue of the fact that it leads to photosynthesis. If photosynthesis were merely a by-product of the adaptive working of chlorophyll, then, as a supposed counter-example to my analysis will demonstrate later in this section, it would hardly be a 'function' of chlorophyll to perform photosynthesis.)

Now where does this leave Nagel's claim that his functional statement also implies (2) (i.e. that plants are goal-directed)? If it does in fact imply (2) then there are four important possibilities open.

(a) The statement implies (1'), (2), and (2'), where these are three logically independent statements.

(b) The statement implies (1') and (2), and further, (2) is equivalent to (2').

(c) The statement implies (1') and (2), and, although (2) and (2') are not equivalent, (2) implies (2').

(d) The statement implies (1') and (2'), and, although (2) and (2') are not equivalent, (2') implies (2).

At this stage of the discussion we can ignore possibilities such as (1') alone implying (2), or (1') and (2') together but not alone implying (2). In itself, the statement that plants use chlorophyll when they perform photosynthesis tells us nothing at all about the goal-directness of plants. However, later, we shall find a class of functional statements where the statement 'z uses x to perform y' does have implications about the goal-directedness of z.

Support for a position not unlike one of these four possibilities is to be found in a recent article by Ayala. Ayala (1968, 218) considers the functional statement 'The function of gills in fishes is respiration'. He agrees with Nagel that such a statement implicitly implies that the particular system being considered (i.e. fishes) is goal-directed; but he also argues that 'in addition it implies that the function exists because it contributes to the reproductive fitness of the organism' (Ayala, 1968, 219). In other words, he believes that the functional statement points to the fact that gills in fishes are adaptations because fish which respire have a better chance of reproducing than fish which do not respire. I am not sure which of the four possibilities Ayala would support, although I rather think it would be (c). He seems to accept Nagel's claim that functional statements imply goal-directedness and then argues that 'Teleological mechanisms and structures in organisms are biological adaptations' (Ayala, 1968, 216).

Of the four possibilities, the canine example seems to rule out (a) and (c) immediately. It was possible to give the sorts of conditions which would lead us to accept the statement 'the function of long hair in dogs is to harbour fleas' as true, without mentioning the goal-directedness of dogs at all. We showed that the peculiarity of a functional statement was that it referred to an adaptation. Hence it would seem that a functional statement can only refer to the goal-directedness of a system if it does so by virtue of the fact that it refers first to an adaptation and then, this in turn, explicitly or implicitly implies goal-directedness. (In rejecting (c) I am saying nothing about the adaptiveness of goal-directed systems in nature. What I do deny is that Nagel's statement implies (2'), by virtue of the fact that it implies (2). This it certainly does not do, because what we would need to infer is not merely that goal-directed systems are adaptive, but that the particular adaptation in question is photosynthesis.) But, given our discussion in the last section, it seems very doubtful that if we say of something that it is an adaptation, we are thereby saying anything about the goal-directedness of the system in which it occurs. Goal-directedness, as we have seen, requires an adaptive *response* given a primary variation. However, if I say of something that it enables its possessor to survive and reproduce (i.e. that it is an adaptation), I do not necessarily say anything at all about what might happen were circumstances to change in any sense, thus triggering a primary variation. Indeed, I might think that were there any significant change, no response would be possible and, purely by virtue of the formally-adaptive character, the chances of survival and reproduction would be practically nil. For example, an organism living in the Arctic might have a white adaptive covering, but be highly (and detrimentally) conspicuous were the background changed from snow and ice to fields and woods. In other words in talking about adaptive advantage and adaptations, one is not necessarily saying anything about goal-directedness.

Therefore neither (b) nor (d) are true. Hence Nagel's functional statement does not imply (2) (that plants are goal-directed).

On the basis of this discussion we are now in a position to state the equivalent non-functional formulation of the generalized functional statement,

The function of x in z is to do y.

It is:

(i) z does y by using x.
(ii) y is an adaptation.

Further, we can also state that (ii) does not imply

(iii) z is goal-directed.

Three possible objections must now be answered. The first objection is as follows. Although it is logically possible that an organism (or group) be in no way goal-directed towards survival and reproduction, as a matter of empirical fact it is highly unlikely that one would ever find an organism or group of such a kind. If it were, then at the slightest environmental change it would probably die (or at least become sterile). Hence, it would seem that every organism and group is in some sense goal-directed. However, since functional statements refer to things which have adaptations, and since the things which have adaptations are organisms, it would seem that functional statements always refer to things which are goal-directed. Consequently, whenever we use a functional statement we imply that the thing to which we refer is goal-directed. The implication is implicit, not explicit; but it is there nevertheless.

This objection rests on a confusion between meaning and reference. It may indeed be the case that every organism is goal-directed; but this fact does not mean that every statement which we use to refer to the organism has a hidden meaning which implies that the organism is goal-directed, and in particular, it does not mean that every functional statement implies that the organism is goal-directed. To draw an analogy, every living man has a heart and brain (persons in the middle of transplant operations excepted); but if I say that man has a brain, this in no way implies that he also has a heart. It is true that he does have a heart; nevertheless, I have said nothing about this. Similarly, an adapted organism will probably be a goal-directed organism; but to say the first does not imply the second.

The second objection to my analysis of functional statements revolves around the way in which I have defined adaptation. A critic might concede that part of the unpacking of a functional statement is

(ii) y is an adaptation;

but then feel unhappy about my claim that saying this is equivalent to saying that

(ii″) y is the sort of thing which helps in survival and (particularly) reproduction,

or, in other words, that (other considerations excluded) an organism with y has a better chance of surviving and reproducing than one without y.

Several people have offered the false statement:

(5) The function of the heart is to produce heart sounds,

as a supposed counter-example to an analysis like mine (e.g. Lehman,

1965b). They feel that if my claim merely were that a functional statement unpacks in part to something like (ii), then they could be happy to agree that since

(6) Heart sounds are an adaptation

is so obviously false, this shows that (5) must be false. But, since my claim is also that (ii) is the same as (ii''), then the complete unpacking of (5) comes out as

(7) The heart produces heart sounds.
(8) An organism (one with a heart) which produces heart sounds has a better chance of surviving and reproducing than one which does not produce heart sounds.

However, conclude the critics, (7) and (8) are true. Hence, it would seem that my analysis commits one to the obviously false, supposedly equivalent, statement (5).

My reply to this criticism is that I do not accept that (8) is true. As things stand, heart sounds just seem to be a by-product of a beating heart, and thus I cannot see that an organism without the heart sounds (but with everything else) would be any less likely to survive and reproduce than an organism with the heart sounds. Of course, this is not to deny that there might come a time when heart sounds would have an adaptive value (and, hence, (5) would be true). Suppose that heart sounds in mammals stimulated infants' sucking. Then it would make sense to assert (5) (obviously, one would still think the heart has other functions).

I suppose that the critic would now object that the idea of an organism with a beating heart but without heart sounds is the idea of something which just could not exist (for physical reasons). However, I think that at this point the critic is being a little harsh on biology. Physics, for example, is allowed to talk about things like frictionless objects (or perfectly elastic bodies), although everybody knows that they cannot exist (for physical reasons). The point is that we know what it is like to go in the direction of complete frictionlessness (or perfect elasticity), and thus, although we can never reach the limits, it seems possible to imagine how something at these limits would be if it could exist. Similarly, the biologist knows what it is like to cut down from really thumping heart sounds to heart sounds which are almost imperceptible. Moreover, doing this seems to make no difference to survival and reproduction. Hence, it seems legitimate to push the idea of low-level-noised hearts to the limit and to say that having no heart sounds at all would make no difference to an organism's reproductive ability, even though we know that an organism without beating-heart sounds is an empirical impossibility.

The third and final objection is, perhaps, more searching than the others. It is obviously rather artificial to talk, as we have done, of an environment E and to suppose that this is some stable, unchanging thing which can be distinguished clearly from all other environments, E', etc. In fact, every environment considered for more than an instant will change somewhat, and organisms within them will have to respond to the changes. Surely, therefore, since adaptations are the things which help organisms survive and reproduce, it is proper to talk of the very mechanisms which help the organisms to respond, as adaptations. Thus, for example, the mechanisms which keep the human body at a constant temperature are adaptations, since a man with his body at a constant temperature has a better chance of surviving and reproducing than one whose body temperature fluctuates wildly. But, in such a situation, does not a functional statement like

(9) The functions of sweating and shivering are to keep the body at a constant temperature

in fact imply implicitly that the system under consideration is goal-directed, for in order to grasp the very notions of sweating and shivering we must understand them as contributing to the goal-directedness of the system of which they are a part. In other words, are there not times when translations along the lines proposed by Nagel are correct?

It must be conceded that insofar as the above argument proves that some functional statements in some sense imply the goal-directedness of the systems under consideration, it is correct. However, it does not prove either that all functional statements imply goal-directedness or, the related point, that translations along the lines suggested by Nagel are correct. To see why this should be so, consider how (9) should be translated. We have

(10) Sweating and shivering help keep bodies at constant temperature,

and

(11) A body's being kept at a constant temperature is an adaptation.

The statements (10) and (11) convey completely the meaning of the functional statement (9). Further, we can infer from (10) and (11) that what is under consideration is a goal-directed system. However, note that we infer this fact from (10), not (11). That is to say, the fact that being at a constant temperature is an adaptation does not allow us to infer that bodies are kept this way by goal-directed systems.

Consequently, considering the generalized functional statement

The function of x in z is to do y,

the only time that we can infer goal-directedness is when the first part of the translation, (i), implies goal-directedness. Obviously in many, if not in most cases, the first part of the translation does not imply goal-directedness (for example neither (1') nor (3) imply it). Hence, most functional statements do not imply goal-directedness. Further, even in those cases where it does imply goal-directedness, Nagel's proposed translation is still incorrect. He would claim that (9) translates as

> (10') Sweating and shivering are necessary for the keeping of bodies at constant temperature;

and

> (11') Bodies are goal-directed (toward survival and reproduction).

For him, as we have seen, the goal-directedness would come in the second part of the translation, whereas, if it is to come anywhere, it must be in the first part. It is easy to see that (10') and (11') cannot possibly be the correct translation of (9) because if it were the case that the fact that bodies are kept at constant temperature makes no difference to the reproductive fitness of men, then even if (10') and (11') were true, (9) would still not be true.

It should be clear by now that Nagel's approach is incorrect. There are indeed goal-directed systems in nature. But when one talks of something having a function, to talk of the very fact of having a function is to say nothing at all about goal-directedness (unless one is specifically talking of the function of the actual goal-directedness of a system). Rather, talk of functions implies talk of abilities to survive and reproduce.[2]

9.3 Teleological explanations

Connected with the two kinds of things we have discussed in this chapter—goal-directed systems and things having functions—one finds two different kinds of explanations, although I shall suggest that only one of these kinds is in any interesting sense a 'teleological explanation'. I shall now look briefly at these two kinds.

[2] I am, of course, restricting my discussion to the modern context. Pre-Darwinians talked of 'functions' and 'adaptations' without the background of a theory of evolution through natural selection. I am not therefore claiming that my analysis is applicable to every instance of function-talk that has ever occurred; but I do think it applies to such talk in modern biology. One who would not accept my connexion between adaptation and function is Munson (1971), who offers three reasons why he thinks such a close link to be untenable. I criticize Munson in Ruse (1972), and Munson (1972) is a reply to me.

In a typical *goal-directed explanation* we consider a system which has actually achieved its goal or which is performing adaptive variations of a kind which would normally lead to a goal. The problem is to explain how the goal or the adaptive variations come about, given an initial disruption (i.e. a primary variation). From a philosophical viewpoint this kind of explanation does not seem very troublesome, and in particular, it does not seem necessary to suppose that such an explanation will be other than a straightforward covering-law explanation. (It must be reiterated, this straightforwardness is one of theory, not of practice. An actual goal-directed explanation might be very complex.)

A good example of the kind of explanation I have in mind here is that which explains the plover's *G*-state, namely the having of a clutch-size of 4. The *explanandum* is the fact that the clutch-size is 4 (or, if the *G*-state is supposed not yet to have been achieved, whatever the average clutch-size might be at the time the explanation is demanded). The *explanans* contains some statements about initial conditions. For example, there will be information about the fact that the clutch-size of some of the birds has been lifted (or reduced) from 4, and together with this will be information to the effect that the change in clutch-size is a function of a genetic change (possibly the normal allele *A* has been changed to a mutant *a*). Also the *explanans* will contain the information that the carriers of the abnormal genes (i.e. *a* rather than *A*) are at a selective disadvantage with respect to the normal gene carriers. Obviously, if the biologist hopes to find just how many generations are needed to reachieve the *G*-state, he will have to attempt to quantify this selective disadvantage. Finally in the *explanans* we will find laws, for example, the Hardy–Weinberg law. Then, from all of these items in the *explanans*, in a manner discussed in earlier chapters, the biologist can hope to infer the way in which the gene-ratio of the abnormal genes (causing abnormal clutch-sizes) declines. His conclusion, the *explanandum*, will either be that the abnormal genes vanish (i.e. the *G*-state is re-achieved) or that, assuming insufficient generations have elapsed for total elimination, that the abnormal genes have been much reduced in number. (More accurately, his ultimate conclusion will be about the phenotypic effects of the vanishing of the abnormal genes.)

This, as I have said, is an unexceptional example of a covering-law explanation, and hence, after one final comment, I shall say little more about this kind of explanation or about goal-directedness in general. My final comment about goal-directedness is that, in a sense, I think the philosophers' concern with goal-directedness has been something of a red-herring as far as the whole topic of biological teleology is concerned. Biological goal-directed systems are, I think, often very complex and very interesting kinds of biological adaptations—they seem to relate to ends in a rather more dramatic way than do non-goal-directed adaptations. However, I suspect that this is all a question of degree, and that in themselves biological goal-directed systems have no kind of even apparent teleology other than that which any functional system has. (Because they are so complex, I doubt whether in nature one does ever get a biological goal-directed system which is not either an adaptation or adapted.) For this reason, disregarding for a moment the functional aspect of goal-directed systems, one can, as I argued in the first section of this chapter, eliminate without trouble any apparent reference to future causes, and for this reason goal-directed explanations cause no special difficulties. Also for this reason it has proven impossible to distinguish between a biological phenomenon like sweating and a non-biological phenomenon like a swinging pendulum, because, questions of function apart, there is no essential difference (see Wimsatt, 1970). Thus, what I would conclude is that in a way the phenomenon of goal-directedness (although usually, if not always, occurring in a functional situation) has hindered the philosopher in his quest for the correct analysis of biological teleology (it certainly has in the specific case of Nagel). If there is some kind of interesting biological teleology to be found in biological goal-directed systems, it is purely by virtue of the fact that, almost inevitably (if not inevitably), they are biologically adaptive systems.[3]

I come now to the second kind of explanation to be discussed, *functional explanation.* This is the kind of explanation in which one tries to explain the existence of some part of a system by reference to its function (or functions). This part of a system could be either a part of an organism (if the organism were the system); but if the system were a group of organisms, then the part could be a whole organism or a subgroup of the whole group. Typical examples of functional explanations are the explanation of the eye by reference to the fact that it serves the function of seeing, the explanation of a woman's breasts by reference to the fact that they serve the function of feeding children, and the explanation of the plover's clutches by reference to the fact that they serve the function of bringing forth young. Obviously, in one explanation one might cite a number of functions an item serves—for example, eyes not only serve the func-

---

[3] Nagel's problems stem from the fact that his analysis draws heavily on Sommerhoff (1950). In that work supposedly we get an analysis of 'adaptedness' in terms of goal-directedness. What we get in fact is an analysis of 'adaptability'—a term biologists use when organisms respond in the face of change. Notice how in the quote on p. 100 Manser uses 'adaptable' when he means, or at least when he should mean, 'adapted'.

tion of seeing, but also, in some cases, of attracting members of the opposite sex.

Now, there are a couple of matters of philosophical interest arising from functional explanations. First, let us look at a charge made by Hempel, namely that functional explanations are usually less than adequate. Hempel tries to spell out in detail the form of a functional explanation, in order to see how closely such an explanation conforms to the covering-law model. Speaking generally of a trait i in a system s (at a certain time t), he offers the following general form of a functional explanation of i.

(a) At t, s functions adequately in a setting of kind c (characterized by specific internal and external conditions)

(b) s functions adequately in a setting of kind c only if a certain necessary condition, n, is satisfied

(c) If trait i were present in s then, as an effect, condition n would be satisfied

---

(d) (Hence,) at t, trait i is present in s (Hempel, 1959)

Hempel points out that, formally, this argument is invalid—in particular, it commits the fallacy of affirming the consequent. The conclusion would follow only if trait i were necessary for condition n. Since it is not, Hempel claims that in no sense can we claim that a functional explanation gives us an adequate covering-law explanation (and hence, Hempel claims that we have no adequate explanation at all).

My own feeling is that Hempel is altogether too severe on functional explanations when he argues in this way. Admittedly, some functional explanations may be inadequate for the reasons he gives; but then, I find it implausible to suppose that any biologist would be satisfied with them. Under normal circumstances I would suggest that when a biologist offers a functional *explanation* of an item, he is in fact supposing that the item is necessary in some sense for the condition which it is causing (although, as we saw in the last section, usually if one says merely that something serves a function, this is to say nothing about its necessity). Hence, I would conclude that, from the viewpoint of the covering-law model, functional explanations create no problems. (The required law in the *explanans* is, of course, (b); but I think that (c) or its substitute would also be law-like, although perhaps just in a loose sense.)

In support of the point I am making, take for example a functional explanation of a cow's udders. The explanation normally offered would be that a cow's udders serve the function of supplying food

G

for the very young. Now, let us spell this explanation out in the style of Hempel, filling out details in the light of our discussion in the last section (but dropping things not pertinent to the discussion, like a reference to a time *t*):

(*A*) Cows are well-adapted (i.e. they have a good chance of surviving and reproducing).

(*B*) Cows are well-adapted only if they have the means to feed their very young.

(*C*) If cows have udders, then they can feed their very young.

(*D*) (Hence) Cows have udders.

Obviously, as it stands, this argument is invalid. But I think that biologists would also hold

(*C'*) Cows can feed their very young only if they have udders.

and then, of course, (*D*) follows without trouble. Naturally, this is not to deny that it is logically possible to find some other way of feeding very young cows. The point is that, as things stand at the moment (and excluding, of course, the interference of man), unless mature cows have udders, baby cows will starve. Thus, in a very real sense, the udders are necessary for the continuation of *Bos taurus*. This being so, I see nothing wrong with offering as an explanation of (*D*), a functional explanation based on (*A*), (*B*), (*C'*).

Moreover, were it the case that cows had an alternative way of feeding their young—say they could regurgitate partly digested food as some birds do in courtship—then I think that if one were to ask for an explanation of a cow's udders, a biologist would take this fact into account. For instance, if cows did in fact have two such elaborate independent ways of feeding their young, presumably there would be some reason for it. Possibly, because cows live in conditions of fluctuating food supplies, they regurgitate in plenty and suckle in famine. In this case, I think the biologist might alter (*B*) to

(*B'*) Cows are well-adapted only if they can feed their very young under all conditions.

His final premise would now become

(*C''*) Cows can feed their very young under all conditions only if they have udders.

Given the fluctuating environment (*C''*) would be true, and (*D*) would continue to follow without trouble. Alternatively, if the biologist could find absolutely no reason for the existence of different ways of feeding the young, then I think he would just drop his original explanation (rather than holding on to its invalid shell as Hempel implies). The biologist would now claim only that he could explain why the cow had feeding methods—not why the cow had the particular methods that it had. In other words, I would suggest that whilst possibly the social scientists are more lax in this matter, Hempel's criticism is not entirely fair to biologists.

The other point I want to make about functional explanations involves the order in which they present cause and effect. Normally, although as we saw at the beginning of this chapter not invariably, when causes and effects are involved the physical sciences explain by putting the causes in the *explanans* and the effects in the *explanandum*, and even in the explanation using Fermat's Principle, we get the cause of the *explanandum*-event in the *explanans*, although admittedly the *explanandum* gives the cause of one of the *explanans*-events. (Frequently in the physical sciences we get arguments with causes in the conclusion and effects in the premises. These are arguments, often called 'retrodictions', where one tries to 'predict in reverse', pinpointing past phenomena from present phenomena and laws. However, in these cases one does not get the understanding associated with explanation.) In a manner similar to explanations to be found in the physical sciences, in goal-directed explanations the cause comes in the *explanans* and the effect in the *explanandum* (which, in the light of my comments earlier in this section, is not surprising). The fitness in the past of birds with 4 eggs, something talked of in the *explanans*, is the cause of the present plover clutch-size being 4, something talked of in the *explanandum*. However, in functional explanations, the positions of cause and effect are reversed. To explain the eyes in terms of their function, seeing, is to explain the cause, the eyes, in terms of the effect, the seeing. Similarly, if one asks for a functional explanation of the plover's eggs, then one does it in terms of what one thinks the eggs will do, not in terms of what went on before. And, of course, more generally in talk of functions we continue to get reference to the future. Even if one adopts my translation of talk of functions into terms of talk of adaptations, one gets reference to the future, since one is now talking of chances of survival and reproduction in the future.

Now, let me re-emphasize that this does not mean that there is any place for future causes in biology—the whole point is that the eyes and the eggs come before and cause the seeing and the offspring. Nevertheless, this order of cause and effect in the explanations (and general reference to the future in function-talk) does, I think, point to something important. This is that, in a sense, there is an irreducible teleological element in biology. This is not the kind of teleology of future causes—they have no more place in biology than they do in physics—but it is a genuine teleology in that we try to understand the world with reference to the future rather than to the past. We try to understand the eyes and the eggs with reference to what they will do, rather than (or, as well as) with reference to what went on before. Possibly one might argue that for an adequate science all one needs is an understanding of the past (and present), and this could be true. However, such an understanding could not include the understanding involving the future (this seems to be a logical point). Hence, I conclude that in a sense (whose significance I shall discuss in the next chapter), since we find it illuminating to consider the organic world with respect to its future as well as its past, biology has an untranslatable teleological flavour distinguishing it from the physical sciences (or at least, from most parts of the physical sciences). But note, this is not necessarily an uneliminable teleological flavour that biology has. One might just replace every functional explanation with a non-teleological explanation—for example, one might explain the udders on present cows by reference to selection on *past* cows, rather than by reference to what one thinks that present-cows' udders will do. In fact, I think this is the kind of move biologists might often be tempted to take. However, it does seem to me that in a case like this, one is not translating a teleological explanation. Rather, one is replacing one's teleological explanation with a different, non-teleological explanation. The teleology itself cannot be translated away. [4]

---

[4] I suspect that the conclusions I have just drawn about teleology in biology will be unpalatable if not unacceptable to many readers. In order to alleviate some doubts, let me point out that in an important sense biological functional explanations are an anachronism—although they are still so common that I would not suggest that any particular user is himself anachronistic in his use. Functional explanations are a carry-over from the pre-Darwinian pre-evolutionary biological era, when the dominant biological paradigm was the Argument from Design (for God's existence). Thus Whewell could write that 'each member and organ not merely produces a certain effect [but] was *intended* to produce the effect; . . . each organ is designed for its appropriate function; . . . each portion of the whole arrangement has its *final cause*; an end to which it is adapted, and in this end, the reason that it is where and what it is' (Whewell, 1840, 2, 79–80). Here it is clear that things are being explained in terms of what we would call their effects— modern evolutionists can and do do the same thing, not because they are committed to Design, but because of the repetitive way in which the selection-process acts, allowing one to suppose for the present and future what happened in past generations, and because of the quasi-Design effect selection has. (See Young, 1971, for a discussion of how pre-Darwinian teleological thought found its way into the *Origin* and later biological works. And see Wimsatt, 1972, for a masterly discussion of biological functional thought—a discussion which I would like to think complements, rather than contradicts, my analysis.)

I O

BIOLOGY AND THE PHYSICAL SCIENCES

Throughout this book we have come up constantly against the question of the extent to which biology is like the physical sciences. Finally, we meet this question head-on, because I want now to discuss implications of the recent, exciting developments in biology which have come from the attempt to bridge the gap between physics and chemistry on the one hand and biology on the other. First, I shall expound briefly the major results claimed in that area of biology which has most felt the onslaught of the ideas and techniques of the physical sciences—genetics. Secondly, I shall consider the relationship between the new 'molecular genetics' and the older, non-molecular, Mendelian genetics. Thirdly and finally, I shall raise the question of whether or not we can look forward to the day when biology as an autonomous discipline will vanish. In particular, I shall consider a number of arguments by 'organismic' biologists—biologists who think there will always be a place for an independent biology.

10.1 Molecular genetics

Unlike Mendelian genetics, one cannot really understand molecular genetics unless one approaches the problems of heredity by first considering the nature of the cell. Now, from a molecular viewpoint, most of the cell seems to be made up of two kinds of things. First one has the structural proteins—these form the building blocks of the cell and go to make up the walls, membranes, and so on. Secondly one has the enzymes, these being proteins which act as catalysts. They enable the cell to go about its chemical activities of breaking down and building up in an orderly manner and at much lower temperatures than such activities would require without their presence. Proteins are polypeptide chains or combinations of such chains, and these in turn are long, string-like molecules made up of literally hundreds of links—each link being an amino acid. Since proteins play so ubiquitous a role in the cell and since they have so many parts, it might seem to be a task beyond human ability to bring even a semblance of understanding to the molecular level of the cell, and hence, to the problems of heredity. However, formidable though the task may be, there are some guides for reason. The chief one at this point is that although there are so many different proteins, each performing such different tasks, the types of amino acid building block are restricted to twenty. It would seem therefore that any difference in protein is a function of a difference in amino acid order. Using this fact as a key, one can now start to unlock some of the molecular secrets of the cell, and one can build up a very detailed molecular theory of genetics.

First, one must locate within the cell the templates which can serve both for the manufacture of fresh supplies of protein and for reproducing themselves, thus passing on to new cells the information required for protein synthesis. It turns out that these templates are not themselves proteins, but are instead nucleic acids. There are two kinds of nucleic acid, deoxyribonucleic acid (DNA) and ribonucleic acid (RNA), and since it is usually the former which is the ultimate carrier of genetic information, let us take a close look at it.

In large organisms DNA is to be found in the chromosomes; but some small organisms dispense with chromosomes and the nucleic acid is to be found on its own within the organism's shell. It is another of the long macromolecules which are so prevalent in the cell, being a polymer of deoxyribose sugars joined by phosphate links. To each sugar is attached, as side chain, a nitrogen-containing base which must be one of four kinds, either adenine or guanine (purines) or thymine or cytosine (pyrimidines). (The bases, together with the sugars and phosphates, are called 'nucleotides'.) The DNA molecule is normally paired, and the two molecules are twisted around each other to form a helix. One finds that adenine on one molecule pairs with thymine on the other, and guanine pairs with cytosine, and it is believed that it is in the order of these four bases along the DNA molecule that is carried the information required to make new proteins.

Now, as we have seen, DNA must do two jobs. The first task is that of replication, so that its information can be passed on to new cells. In order for this to happen the two strands of DNA in a helix start to come apart, and then, with the aid of enzymes, complementary nucleotides line up against the nucleotides on the unzipped single strands of DNA. Thus the precise order of the nucleotides is passed on to a new DNA strand, although obviously the new strand is the complement of the old strand and it must therefore duplicate itself before one gets a DNA strand identical to the first strand. The second task of DNA is to make proteins—it does not do this directly but via an intermediary, RNA. RNA is like DNA in that it is a long strand of nucleotides, but where DNA has thymine, RNA has uracil. There seem to be three types of RNA, all with different functions within the cell: messenger RNA (mRNA), ribosome RNA (rRNA), and soluble or transfer RNA (sRNA or tRNA). All three types are copied off DNA in much the same way as DNA replicates itself, except that a different enzyme is involved, and obviously uracil (not thymine) pairs with adenine. The rRNA molecules, together with some proteins, go to form the ribosomes, which in turn serve as the sites of protein synthesis within the cell. The mRNA travels from the nucleus to the ribosomes, carrying with it the information needed for the synthesis of different proteins. Finally, the sRNA picks up free amino acids within the cell, bringing them over to the ribosomes, where they can be lined up in appropriate order against the mRNA, joined and cast off as completed polypeptide chains. Thus one gets the all-important synthesis of proteins.

Clearly, this is but a sketch of DNA replication and of protein synthesis. Necessarily many important steps have been omitted. One omission which must be remedied arises from the fact that proteins consist of twenty amino acids, but that RNA (like DNA) carries but four different nucleotides. This means that there cannot be a 1–1 correspondence between the nucleotides of mRNA and the amino acids of proteins. It is suggested that three nucleotides code for an amino acid, and since there are 4^3 possible different triplets of ordered nucleotides, it is also suggested that many of the triplets are degenerate (i.e. code for the same amino acid) and that other triplets make no sense at all (i.e. do not code for any acids). Recently, molecular biologists have made great advances in finding exactly which triplets code for which amino acids.

So far in the discussion we have assumed that everything always goes according to plan—however, even in the molecular world things can go wrong, and so let us now see how normal protein synthesis might be disrupted. The most obvious place at which trouble could start is on the DNA molecule—if something happens here then not only is it liable to be reflected right through to the proteins of the present cell, but also it is likely to be passed on to new cells as well. Three types of alterations to the DNA information chain seem possible—nucleotides might be added to the chain, they might be eliminated, and the original nucleotides might be altered. It would appear that all three of these types of changes occur; however, fortunately, it also appears that in all three cases, the overall effect may not be as drastic as one might fear. The extent to which the cell's operations are disrupted depend very much on the particular circumstances. Consider first the addition or deletion of nucleotides. As was mentioned immediately above, the nucleic acid code is read in groups of three, hence, the addition or deletion of one or two nucleotides is liable to be highly disruptive, especially if such an alteration occurs at the beginning of a chain. However, the addition or deletion of a group of three (or multiple of three) nucleotides leaves all (or most) of the original code intact, particularly if the three added (or deleted) nucleotides are very close together. As one might expect, it transpires that codes altered in ways like this are often able to produce proteins which function normally or with only slightly reduced powers. A similar situation applies to changes of the code which take place by the alteration of simple nucleotides. There are several different things which might happen. First, the alteration of one nucleotide might change the triplet of which it is part to a triplet which codes for the same acid, in which case the protein would be unchanged. Secondly, it might change its triplet to the code for a different acid (this is called a 'mis-sense mutation')—here we could get a protein very similar to the normal protein with some or all of the capabilities of the normal protein. Thirdly, the change might produce a triplet with no corresponding amino acid (a 'nonsense mutation'). Since it is thought that a triplet which does not make sense can end the synthesis of the protein polypeptide chain, presumably the extent to which a nonsense mutation would render the protein non-functional would depend in part on how close to the end of the chain the nonsense mutation occurred.

It is clear, therefore, that changes in the DNA code can occur in several ways and have several effects. It can leave the protein synthesis unchanged, it can reduce the effectiveness of the protein, it can change its function, it can stop its synthesis altogether, or it can change the protein back to an original form—it should be noted that in the latter case, the change of the protein back to an original form (or at least to an original function) does not necessarily mean that the nucleic acid sequence is back in its original form.

Finally in this section, let us turn our attention to the phenomenon of crossing-over, something which has been found to occur, not just between chromosomes, but between naked strips of DNA. Although its ultimate cause is still unknown, a good deal of light has been thrown upon it. Of particular importance to us here is the fact that molecular biologists have been able to show that crossing-over can occur between many, if not most of the base pairs of a DNA molecule —if this is the case, then it would seem that the smallest unit of crossing-over (i.e. the unit within which crossing-over is impossible) is probably a single pair of (matched) nucleotides. As we shall see in the next section, this fact leads to consequences of great philosophical interest.

A lot more could be written about the findings of molecular cell biology; however, I think that enough biological territory has now been covered for our purposes. Let us, therefore, now turn to a discussion about the relationship between this molecular genetics and the non-molecular genetics we encountered in the early chapters of this book. (Watson, 1965, 2nd ed. 1970, is a good introduction to molecular genetics.)

10.2 *The relationship between molecular genetics and Mendelian genetics*

Many philosophers argue that although a new theory may displace its predecessors entirely, often there is a continuity between the old and the new. This is by virtue of the fact that older theories can be shown to be special instances derivable from the more general premises of the new theory. Recently, however, a number of philosophers have argued that this 'reduction' of one theory to another is a philosophical fiction. They argue that theory-change always involves the complete rejection of the older theory (from the view-point of truth). In order to discover the true relationship between molecular genetics and Mendelian genetics, I shall begin this section by considering in some more detail some of the arguments in this debate about theory-reduction. Having done this, I think I shall be able to throw some interesting light on the relationship between the two genetics.

Let us ask first what exactly a reduction of one theory to another would involve. Ernest Nagel, a strong believer in the possibility of theory-reduction, argues that if we are to speak of one theory being reduced to another then what this must mean is that we can show the one theory to be a *deductive* consequence of the other. We must be able to show that all of the propositions of the reduced theory can be deduced from axioms of the reducing theory. Now, Nagel admits that sometimes the reduced theory will contain terms not occurring in the reducing theory.

Accordingly, when the laws of the secondary science do contain some term '*A*' that is absent from the theoretical assumptions of the primary science, there are two necessary formal conditions for the reduction of the former to the later: (1) Assumptions of some kind must be introduced which postulate suitable relations between whatever is signified by '*A*' and traits represented by theoretical terms already present in the primary science . . . (2) With the help of these additional assumptions, all the laws of the secondary science, including those containing the term '*A*' must be

H

logically derivable from the theoretical premises and their associated co-ordinating definitions in the primary discipline. (Nagel, 1961, 353–4)

Nagel calls the conditions (1) and (2) the condition of connectability and of derivability respectively, and he is at pains to note that obviously we must have some reasons for thinking it proper to accept the inter-theoretic links demanded by the condition of connectability. We would not, for example, think much of a reduction which linked cabbages in one theory with molecules in another.

Two objectors to the whole idea of theory-reduction are Kuhn (1962) and Feyerabend (1962), and they base their opposition essentially on historical grounds. The trouble for them seems to be not so much that there is anything formally wrong with what Nagel writes, but rather that, as a matter of historical fact, one has never had a reduction of one mature scientific theory to another scientific theory in the way that Nagel has just described. That is, they claim that one never finds a deductive connection between a theory and its successor. An example that Feyerabend uses to support his case is that of the supposed reduction of Galilean physics to Newtonian physics. In fact, Feyerabend points out, what we can deduce from Newtonian physics are things very close to Galileo's laws of motion, *but not the exact laws*. So long as the ratio of H/R (where H is the height above ground level of the processes being described and R is the radius of the earth) is finite, Galileo's laws will not follow deductively from Newton's laws. But obviously H/R is always finite, and therefore Feyerabend concludes that a Nagelian-type reduction is impossible. Kuhn's example is that of the supposed reduction of Newtonian physics to Einsteinian physics. Kuhn argues, in a manner similar to Feyerabend, that here also we have no deductive relationship. Even if we consider only low velocities where some of Einstein's laws seem approximately similar to Newtonian laws, such similarity is apparent. The concepts in the two laws are quite different. For example, 'Newtonian mass is conserved; Einsteinian is convertible with energy' (Kuhn, 1962, 101). Hence, Kuhn like Feyerabend, generalizing on the basis of a few historical examples, concludes that one never gets a reduction of the type described by Nagel.

Turning now back to biology, let us ask ourselves to which of these two extreme positions the relationship between non-molecular and molecular biology seems best to approximate. First, can a case be made for supposing that one has here a reductive relationship of the type described by Nagel? Now obviously, if we are to argue for a reduction, then one notable omission must be amended. In the discussion in the last section, no mention was made of Mendelian genetics most central concept—the gene. Before we can have reduction, we must have some links enabling us to go from talk at the molecular level to talk about the Mendelian gene.

It is fairly obvious that if such links are to be found, at one end of them will be talk about DNA molecules, and at the other end, talk about Mendelian genes. As we have seen, DNA molecules serve in some sense as the ultimate causes of cell characters, as the units of heredity, as the sources of new variation, as the entities which cross-over, just as we have seen Mendelian genes serving as the ultimate causes of phenotypic characters, and so on. The major question is just *how much* DNA is to be postulated as being equivalent to a Mendelian gene. Again, there seems to be a fairly obvious answer. As we saw in the second chapter, the defining feature of the Mendelian gene is that it is the unit of function, that is, it is the unit ultimately responsible for the organism's contribution to phenotypic characters. Hence, it seems best to suggest that the Mendelian gene be considered equivalent to just sufficient DNA to act as a unit of function, although it should be remembered that when we speak of a molecule of DNA acting as a unit of function, we (usually) mean something slightly different from what we mean when we speak of a Mendelian gene acting as a unit of function. To speak of DNA *qua* function is to speak of something acting as a cause of polypeptide chains—to speak of a Mendelian gene *qua* function is to speak of something acting as a cause of phenotypic characters. Sometimes these two things coincide—after all, the inside of the cell is part of the phenotype—however, there need be no such coincidence. There is a difference between blue eyes and the polypeptides causing them.

It is perhaps worth mentioning that molecular biologists have in fact appropriated the term 'gene', and do now include in their meaning of it a piece of DNA which is just enough to serve as the cause of a cellular product (i.e. a polypeptide chain). This being so, let us next ask whether, if we identify particular molecular genes with particular Mendelian genes, we get a reduction of a kind described by Nagel? Prima facie one might think that one would. Part of the reduction would involve trying to satisfy the condition of derivability, and this, in turn, would require one to deduce from molecular premises laws of non-molecular cell biology, for example, Mendel's laws. Now, the attempt to get Mendel's first law from molecular premises looks very promising. The law demands that the units of inheritance be the units of function. This holds true of molecular genes. The law demands that the units of inheritance be particulate, not blending irretrievably in each generation. This holds true of molecular genes (at least in the sense that there is no irretrievable blending of nucleotides). The law demands that the units of inheritance be distributed in a particular way from one generation to the next. Again there is no reason why molecular genes should fail to satisfy the law in this any more than Mendelian genes. Hence, at first sight, the chances of a reduction seem fairly good. One starts with the relevant statements about molecular genes (e.g. that such genes are particulate), one adds the connecting links between molecular genes and Mendelian genes, one concludes with some propositions about polypeptide chains being phenotypic characters (or the causes of such characters), and one then deduces something remarkably like Mendel's first law —a key premise of Mendelian genetics. (Note that it is not necessary for a reduction that every molecular gene have a Mendelian equivalent. What we need, assuming that we are trying to reduce laws referring to all Mendelian genes, is that every Mendelian gene have a molecular equivalent. Then suppose that 'Mx' stands for 'x is a Mendelian gene', 'mx' stands for 'x is a molecular gene (of a type including all of those with Mendelian equivalents)', 'px stands for x is some specified kind of molecular phenomenon', and 'Px' stands for 'x is some specified kind of biological phenomenon'. Suppose also that we have the molecular law $(x) (mx \rightarrow px)$, i.e. 'All m's are p's', or, 'All m's produce p's', and that $(x) (px \rightarrow Px)$ also holds, i.e. 'the molecular phenomenon p entails or causes the biological phenomenon P'. Then, by virtue of the fact that $(x) (Mx \rightarrow mx)$, we can go deductively from the molecular law $(x) (mx \rightarrow px)$ to the biological law $(x) (Mx \rightarrow Px)$. Apparently a reduction of Mendel's first law can be made to fit this general *schema*, although probably this theoretical analysis is simpler than the practice might be in the cases of some derivations. One might for instance, have certain Mendelian genes equivalent to a combination of molecular genes.)

Unfortunately, I think one can show this apparent reductive success of Mendel's first law to molecular genetics to be illusory. When we come to consider such things as mutation and, more particularly, crossing-over, we see that the two genetics come into conflict, and consequently a reduction along the lines suggested by Nagel seems impossible. Non-molecular genetics supposes that the unit of function, the unit of mutation, and the (smallest) unit of crossing-over, are all one and the same thing. Molecular genetics separates out these units—in particular, it allows that mutation and crossing-over can involve but a very small part of the molecular gene, no more than one or a few nucleotides (of which there can be hundreds in the molecular gene). But this means that, although molecular and Mendelian genes share many properties, none of the genes themselves can any longer be identified with or considered equivalent to each other. Hence, the derivation of Mendel's first law from molecular biology in the way sketched above is illegitimate, and obviously, hope of getting it or other laws by alternative means seems remote if not missing entirely. For example, the revised version of Mendel's second law allows crossing over between but not within units of function on the same chromosome. A molecular equivalent to this law would have to allow crossing-over within units of function on the same chromosome, and thus one could hardly hope to derive Mendel's second law from the closest molecular equivalent.[1]

It seems, therefore, that the pendulum has swung completely away from Nagel—he is wrong (or he would be if he were to suppose that we have a reductive relationship here) and Kuhn and Feyerabend are quite right in denying the existence of such relationships. Actually, however, the true situation is rather more complex than this. Long before 1953, the year in which Watson and Crick discovered the structure of the DNA molecule and made positive suggestions about its role in heredity, biologists had suspected (for biological reasons) that the Mendelian gene concept, as I have so far discussed it, is inadequate. This gene concept, often known as the 'classical gene concept', treats genes like beads on a string. But, as far back as 1925, Sturtevant discovered that the classical gene concept fails to account for every facet of the complex story of heredity. In particular, Sturtevant (1925) discovered the so-called 'position effect', namely that the order of the genes on a chromosome can affect their phenotypic result. Specifically, Sturtevant discovered that two identical genes being on the same chromosome (next to each other) and heterozygous to wild-type genes can have a stronger effect than the two genes being homozygously opposite each other on different chromosomes. In the years succeeding, biologists continued to turn up evidence that the Mendelian gene concept as so far discussed fails to account for all the niceties of inheritance, and then, in the 1950s, thanks particularly to so-called 'fine structure genetics' the old classical gene concept fell entirely apart, and it was replaced by a number of more exact concepts. Let us see how this happened. (See also Carlson, 1966; Whitehouse, 1965.)

Since Mendel, geneticists have sought the ideal organisms with which to experiment. What one needs are organisms which reproduce rapidly, which have many offspring (thus giving rise to a high proportion of mutations and different types of crossing-over), and which are such that one can identify new mutants or peculiar

---

[1] Consistency between two theories is a necessary condition for the reduction of one to the other, although it is clearly not a sufficient condition. Of course, when one has two theories which are couched in different terms, as we do here, one might try to avoid inconsistency, thus permitting a possible reduction, by trying out different connecting links between the theories. But I do not see in this case what links one might substitute for those suggested, namely between molecular and Mendelian genes. And even if one did find alternatives, there would still be the question of why one should accept them as true.

results of crossing-over rapidly. *Drosophila* (fruit-flies), the classical geneticist's favourite organism, represent a great advance over Mendel's pea-plants; but neither fruit-flies nor pea-plants can in any way compare with the organisms that Benzer (1962) and others started to use, namely bacteria (e.g. *Escherichia coli*) and the viruses (often called 'phages') which attack these bacteria. Phages are little more than shells containing naked strips of nucleic acid, they reproduce very rapidly, they are haploid (i.e. one has no trouble with dominant genes masking other genes), and by controlling the media in which they and their bacteria are allowed to grow, one can detect new mutants and rare recombinations (due to crossing-over) from amongst millions of less interesting fellow organisms.

Through using organisms and techniques like these, Benzer and others were able to perform a fine-structure analysis on the biological unit of function, and they rapidly showed that this unit is not synonymous with the unit of mutation or the smallest unit of crossing-over. Mutation and crossing-over can occur within the unit of function, and can indeed involve very small parts of the whole unit. Benzer therefore suggested dividing the classical Mendelian gene concept into three—the unit of function which he called the 'cistron', the unit of mutation which he called the 'muton', and the smallest unit of crossing-over which he called the 'recon'. Other geneticists have accepted this division of the old concept (amongst other things it explains the position effect), and the important thing to notice is that Benzer was able to make the division on the basis of experiments using very sophisticated but *purely biological* techniques of breeding and checking for mutation and crossing-over.

Now, if we consider not the older classical Mendelian genetics but this new yet still entirely biological, fine-structure genetics, then the task of showing that one can have a Nagelian-type reductive relationship between molecular and non-molecular genetics seems once again to be within the realm of possibility. One links the (biological) cistron with the (non-biological) molecular gene and the (biological) muton and the (biological) recon with a very few non-biological nucleotide pairs of the DNA molecule. (Actually, Benzer concluded that the muton and recon can be less than 5 such pairs.) Then, with connections like these between the biological and non-biological, one can hope to satisfy the condition of connectability, and, in turn, the condition of derivability, for we no longer have the conflicts which we had previously. As Benzer (1962) writes, 'everything that we have learned about the genetic fine structure of T4 phage is compatible with the Watson–Crick model of the DNA molecule'. Thus, for example, both of Mendel's laws (which now, at the biological level, talk about cistrons rather than classical genes) can be derived from molecular premises in conjunction with these links between the biological and the molecular. (Obviously, Mendel's second law has to be revised to allow for crossing-over within cistrons; but this is a biological revision.) The incompatibility between the two genetics has vanished, and the one can (in principle) be shown to be a consequence of the other. (Necessarily in this discussion, a number of simplifying, but I think not entirely falsifying, assumptions have been made. I have, for example, totally ignored any complicating factors due to sexuality.)

As a consequence, what I would suggest is that probably we do here have a situation akin to that described by Nagel, although not even genetics is so rigorously formalized as Nagel supposes every science to be and I must confess that strictly speaking I think the present situation is as much that there are now no theoretical barriers in the way of a Nagelian-type reduction and that there are obvious signposts about how this should be done, as that such a reduction has been rigorously accomplished. (As I pointed out, apart from anything else although the elimination of conflict is a necessary condition for a reduction it is not a sufficient condition.) But I would also suggest that the recent history of genetics shows that, were one to claim that the reduced theory is always entirely completed before the arrival of the reducing theory, one would be wrong. What we seem to have had here is both reduced and reducing theories developing simultaneously (and obviously not entirely in isolation). It is only after this development that the physico-chemical and the biological came into harmony, opening the way for a reduction, or at least, for a possible reduction.[2]

10.3 *The organismic position*

Obviously, even if what I have argued in the last section is well-taken, this is a far cry from being able to say that the whole of biology is now or ever could be part of physics and chemistry. Even though genetics plays an absolutely pivotal role in biology, there is far, far

---

[2] It is perhaps important to emphasize that no actual deduction from a purely molecular theory to a purely biological theory seems yet in existence. Indeed, reading the works of biologists one gets the feeling that the whole question of reduction is more of a philosopher's problem that a scientist's. Certainly, biologists do not keep molecular and biological studies rigorously separate, with an eye to showing deductive links between them. Rather, there seems to be a fairly casual merging of both molecular and biological concepts. Hence, biologists like Dobzhansky (e.g. 1970) seem to feel no qualms about introducing talk of genes *qua* DNA segments into their essentially biological discussions. And it is probably for this reason that many biologists have not bothered to adopt Benzer's suggested concepts of cistron, muton, and recon, but rather still talk in terms of 'genes', and refer directly to the structure of DNA when they want to make the distinctions that Benzer's concepts are intended to reflect.

more to biology than just this one theory. Hence, on the basis of the discussion so far we cannot as yet answer the question most frequently asked about biology, namely, will the day come when biology will lose its autonomous status and become just a branch of the physical sciences? It is to this question, in the hope of throwing some light on the problem it poses, that this section is devoted. (See also Schaffner, 1967b.)

Now, in order to keep a sense of proportion, let us start the discussion by acknowledging the achievements so far recorded by molecular biologists—achievements which have actually brought areas previously biological into the sphere of the physical sciences. Even discounting the basic parts of genetics, molecular studies of biology progress at a great rate and in a highly fruitful manner. One area, for example, where the application of the ideas and techniques of the physical sciences has paid great profits is that normally called 'bioenergetics'—the part of biology which is concerned with the flows of energy through living systems. It has long been known that the ultimate source of biological energy is sunlight captured by green plants; but by the application of the ideas of physics and chemistry, in particular by the use of the second law of thermodynamics, the exact way in which such energy is transformed and used is now being revealed. Perhaps the most exciting aspect of bioenergetics is the fact that, in large measure because of the study of energy flow, cell parts like mitochondria and chloroplasts are being analysed so minutely that their very structures and fabrics are being understood in molecular terms—the materials which make up organisms need no longer be considered, in some sense, irreducibly biological, inert containers wherein chemical reactions occur; but are now being revealed through their molecular constitution as being intimate and vital parts of the whole process.

Nevertheless, notwithstanding the great molecular successes in genetics, bioenergetics, and other areas of biology, it cannot be denied that we are still very far from a complete physico-chemical understanding of the whole spectrum of biological phenomena. Even if we ignore the things studied by paleontologists, systematists, and so on, the fact remains that at the most basic levels there are problems which will tax the ingenuity of molecular biologists for years to come. Indeed, Watson writes of *Escherichia coli*, the organism best understood of all at the molecular level, that

[some] cellular molecules, in particular the proteins and nucleic acids are very large, and even today their chemical structures are immensely difficult to unravel. Most of these macromolecules are not being actively studied, since their overwhelming complexity has forced chemists to concentrate on relatively few of them. Thus we must immediately admit

that the structure of a cell will never be understood in the same way as that of water or glucose molecules. Not only will the exact structure of most macromolecules remain unsolved, but their relative locations within cells can be only vaguely known. (Watson, 1965)

In the light of this impossibility from a *practical* viewpoint of ever providing a complete physico-chemical analysis of the organic world (at least in the foreseeable future), the question now arises of whether such a complete physico-chemical analysis is ever possible even from a *theoretical* viewpoint. Obviously, in some sense, one can give a physico-chemical analysis of the organic world—a falling body obeys Galileo's laws whether it be a stone or an elephant—but the question still remains of whether or not physics and chemistry can account adequately for all of the aspects of the organic world which have hitherto been considered peculiarly biological. The sorts of aspects, like the development of organisms, which up to now have been explained by biological theories rather than physio-chemical theories.

In the past, many philosophers (and not a few biologists) argued that such a complete take-over by the physical sciences is impossible, and they based their opposition on the supposition that denizens of the living world are distinguishable from non-living things by virtue of the fact that the former, unlike the latter, contain certain 'life forces'. The possession of such a life force, or (as it would have been called by Bergson) an *élan vitale*, supposedly makes a living thing what it is—something distinguishable from non-living things. Moreover, it was argued that no purely physico-chemical theory could adequately catch the 'true essence' of living things, since no physico-chemical theory could analyse these life-forces. Today, few biologists can be found to give any support to vitalism, that is, to a belief in such forces. The reason for this is not so much that a belief in such forces is contradictory, but rather that their existence or non-existence seems totally irrelevant to the biological endeavour. The forces are indetectable, they are not subject to experimental control, and everything they were invoked to explain seems entirely explicable in some other way—a way not involving a commitment to the forces.

However, even though vitalism seems to be a thing of the past, many biologists still argue that there will always be a place for an independent biology. They feel that some facets of the biological world will forever elude a purely physico-chemical analysis. It is the arguments of these people who commonly call themselves 'organismic biologists' that I want now to consider. I shall take five such arguments. The first two, I think, have little or no merit. The next three have, I think, some force, although I do not think that they prove quite so much as some of their proponents seem to think they prove.

Possibly the most common argument from those who would argue that biology must always remain autonomous is that which bases its claim on the supposed *uniqueness* of biological phenomena. It is argued that the things the biologist studies—cells, organisms, groups —are all unique, to an extent unknown in the physical sciences. But, it is argued, the physical sciences can deal only with repeatable, non-unique phenomena, and thus it is concluded that the science dealing with biological data cannot ever be entirely physico-chemical. For example, Bentley Glass, one supporter of this argument, writes that

the uniqueness of the particular event, embedded in the history and evolution of life, seems an unanswerable argument for the possibility of explaining all aspects of life in terms of the laws of physical science which are demonstrable in non-living systems. (Glass, 1963, 248)

In an earlier chapter, I pointed out that much of the force of this argument rests on a blurring of several senses of 'unique'. Admittedly, in one sense biological phenomena are unique, and admittedly, in a sense, physico-chemical theories cannot deal with the unique; but the uniqueness involved in these two cases is of different sorts. Looking at the question in one direction, everything is unique—if nothing else, everything has its own spatio-temporal coordinates. The point is that physical scientists abstract, and they consider only properties shared by more than one thing. There seems to be no reason why they should not do the same thing with biological phenomena. After all, a fact frequently overlooked by organismic biologists is that biologists themselves abstract from the unique and consider non-unique aspects. Consider, for example, the entirely biological classical Mendelian gene concept. One striking feature about such a gene is that a particular instance can be repeated an infinite number of times, and all of these instances are *absolutely identical*. If I pass on a blue-eye-causing gene to my son, then, barring mutation, classical Mendelian genetics postulates that instances of my gene and instances of his gene are the same kind. They are not more or less alike; rather, just like two hydrogen atoms, they are indistinguishable (even to God). As we have seen, paradoxically, the molecular biologist would distinguish genes not distinguished by the classical geneticist (but distinguished by the fine-structure geneticist). Consequently, I would suggest that, popular though the organismic arguments based on biological uniqueness may be, in fact the uniqueness of biological phenomena does not bar the complete molecularization of biology. On the one hand, even as things stand at the moment, physics and chemistry can deal with unique phenomena (through abstraction). On the other hand, biology itself has to abstract from unique things and deal with common regularities.

Another argument popular with organismic biologists is that, just as in quantum physics so also in biology, one must allow for a *principle of complementarity*. It is argued that biological laws are framed for intact living organisms, but that to find out things about organisms at the molecular level, it is necessary to rip the organism apart, thereby (usually) killing it. Thus it is claimed, just as in physics, the closer one comes to finding out one thing, the more likely one is to distort something else—in particular, there may be peculiarly biological phenomena which the methods of molecular biology necessarily conceal (see Delbrück, 1949).

Again it is not too difficult to show that this argument does not carry much weight. Its great weakness is that it is only an analogy from physics, and whilst analogies *per se* are not bad things, I think arguments need to be given to show that it is needed in biology. In the absence of any such compelling arguments, one is left very much with the feeling that applying the complementarity principle in this context is an *ad hoc* device to save one's prior belief in the autonomy of biology, not something which of itself compels one to accept the autonomy of biology. Hence, until such time as phenomena are encountered which can be explained only by some kind of biological complementarity principle, I see little reason to take this argument seriously. On top of this, the argument seems to rebound to a certain extent on non-molecular biology. Non-molecular biologists often kill their subjects before studying them —indeed paleontologists never deal with live subjects—and even when the subjects are left alive, often severe alterations are made in their modes of living so that they probably do not behave as they do when unobserved (for example, as we have seen, some organisms of the same species refuse to breed in captivity). Thus, given a complementarity principle, one could argue that non-molecular biologists stand in as much danger as molecular biologists of losing sight of those things which make biology autonomous.

We come now to three arguments which, I think, have a little more weight than those just discussed. (All three occur in Simpson, 1963b.) The first argument revolves around the supposedly *historical* nature of organic phenomena. We have already seen some aspects of this argument at work in our discussion of evolutionary theory, and I have argued that some of the claims made on the basis of the supposedly historical nature of biological phenomena are illegitimate. In particular, I have argued that one must draw a distinction between an evolutionary theory and a particular phylogenetic path taken by a group of organisms. Unless one does this, one is liable to ascribe to one's theory an historical dimension proper only to phylogenies. Nevertheless, this having been said, it cannot be denied that biological concepts commonly have an historical element not to be found in the concepts of physics and chemistry. Consider, for example, the concept 'water'. To say of something that it is water is (normally) to say nothing at all about its history. We do not know, for example, if it was obtained by burning hydrogen in oxygen, or methane in oxygen, or in some other way. On the other hand, to say of something that it is, for example, a 'zygote' is to tell us something of its history. It is a cell caused by the fusion of two other cells, which latter come from two different organisms. If it did not come about this way then it is not a zygote. Similarly, at least under Simpson's understanding of taxonomy, to say of something that it is a member of *Homo sapiens* is to tell us something about its history. Amongst other things, it had to be born of another member of *Homo sapiens*, otherwise it could not be a member of *Homo sapiens*. (Obviously, the earliest members of the group are excused from this demand.) Hence, it would seem that biology does try to account for an historical dimension in the things that it analyses, in a way not to be found in physics and chemistry.

The question to be asked is what exactly this proves. Does it, for example, mean that one can never, even in theory, have an adequate physico-chemical analysis of, say, evolutionary phenomena? I rather think not. For a start, it is not strictly true to say that the concepts of physics and chemistry never have an eye towards the past. Consider for example the phenomenon known as 'hysteresis'. The flux density (roughly speaking, the magnetic power) of a Rowland ring of ferromagnetic material (this is a ring with wire coiled around it and with current through the wire) is a function not only of the magnetic intensity at a particular time (basically the current flowing at a particular time) but also of the *past history* of the ring. To be told something of the present flux density and magnetic intensity can be to be told something of the ring's past. Hence, in an analogous way, one might in the future have a theory dealing with the historical aspects of biological phenomena and yet, on the basis of today's physics and chemistry, still be prepared to call it 'physico-chemical' —there seems to be no logical reason why a physico-chemical account must be non-historical.

A complement to this argument is another, based on the fact that not even biologists seem to consider the historical content of their concepts to be something to be preserved at any cost. We have seen how, already, some taxonomists seem to read far less of an historical aspect into their taxa than do others, and it would seem that even with something like the zygote, the historical aspect could be dropped. What would be required instead would be a biochemical-cum-morphological description of the cells in question, distinguishing them from other cells. Were such a description provided, then I do not think biologists would feel particularly cheated out of their old concept, particularly if, for example, one could artificially create zygote-like cells, indistinguishable from naturally produced zygotes. Hence, in a manner similar to that discussed in the last paragraph, one might in the future have a theory considering only the non-historical aspects of biological phenomena, and yet, on the basis of today's biology, still be prepared to consider it adequate. (I try to make this point in a rather different context in 1970b.)

Actually, I think what one might expect to find is a meeting of today's fairly non-historical physics and chemistry and today's fairly historical biology, where the concepts are half historical and half not. It seems to me that something like our thinking about DNA already occupies this middle ground. The DNA molecule is obviously in one sense purely chemical and perhaps in this sense non-historical. However, no DNA molecule has yet been synthesized entirely without the aid of other such molecules. Hence, to be told that something is DNA is, in another sense, to be told something about its past, namely that another DNA molecule was involved in its making.

Nevertheless, whilst all the points I have been making seem to me to be true and to take the major sting out of the organismic biologist's argument for the autonomy of biology on the basis of the historical dimension of biological phenomena, one aspect of this argument does still seem to hold true. This is that, as things stand today, given the present essentially non-historical physics and chemistry and the present essentially historical biology, the absorption of the latter into the former seems unlikely, if not impossible. Until such time as one side or the other changes its attitude about the relevance of the past to the concepts of science, major parts of biology would seem to be destined to stay autonomous.

The second significant argument put forward by organismic biology's supporters is, more or less, a converse of the argument we have just been considering. It is the argument that biology looks to the future, particularly in its concern with *functions*, in a way that the physical sciences do not. Morton Beckner, for example, argues that if we speak of something serving a particular function, then we are certainly not in any sense committing ourselves to a doctrine of final causes. However, he does feel that we are directing our interest towards the future in a manner which would be lost if we were to try to eliminate our concern with functions entirely. Beckner makes his point with the following example:

> Suppose we are watching a tank in which there is a single anchovy, and that we introduce a barracuda into the tank. At first nothing happens; then, when it would be reasonable for us to suppose that the anchovy spots the barracuda, the anchovy behaves as follows: he turns sharply, swims quickly to the surface, leaps out of the water, reenters, and repeats the sequence. This whole description (call it B) is, let us at least suppose, in non-teleological language. And it is a description of exactly the same behaviour that would serve as the basis for the teleological description (call it A) 'On sighting the barracuda, the anchovy engaged in an escape reaction.' Of course B does not translate A. A in some ways says less, and in other ways says (and presupposes) much more. A says less than B since B offers details not mentioned in A. It says more, since calling the anchovy's behavior an 'escape reaction' implies that it serves the function of escape, whereas this is not implied by B. (Beckner, 1969, 162–3)

Beckner's feeling, therefore, is that an entirely non-functional treatment, which is all present physics and chemistry could offer, would be bound to lose something that the biological functional treatment possesses. In this I agree with him (as I think my discussion at the end of the last chapter indicated), and hence I would want to conclude in much the same way I concluded my discussion of the historical nature of biology. One could have a non-teleological non-functional treatment of biological phenomena, in particular, one could have a physico-chemical treatment of biological phenomena, and I see no reason why this should not be entirely adequate. However, for better or for worse, such a treatment of biological phenomena would not do everything done by present-day biology. In particular, it would not direct our attention to certain things, in the light of what we would think will happen to them in the future. One can eliminate the teleology but one cannot translate it away. Alternatively, possibly one could introduce such a concern for the future into one's physics and chemistry, and the explanation using Fermat's Principle seems to show that this would not be logically impossible; but then again, one would still not be explaining every aspect of the biological world given the physico-chemical theories that we have today. Therefore, either way one looks at the problem, it would seem that there are aspects of present-day biology which elude present-day physics and chemistry. Hence, as before, there is some truth in this organismic argument.

The third and final organismic argument which I think has merit revolves around the nature of biological *order*, although, once again, I do not think as much is proved as some organismic biologists think. There can be no doubt that biological phenomena, from the genes on the chromosome to the members of the species, are highly organized. Many writers have seized upon this fact and argue that biological theories can grapple with and account for this order; but they believe that it provides too great a challenge to the theories of physics and chemistry. Hence, again they conclude that biology must always remain autonomous. (See Polanyi, 1968; but see also Causey, 1969.)

Obviously, in an extreme form this argument fails. One certainly cannot claim that physics and chemistry pay no attention at all to order. Indeed, order can make a vital difference. Consider, for example, the laws governing currents going through resistors. If one has two resistors (magnitudes R_1 and R_2) and they are put in series, then their combined resistance is given by the law:

$$R = R_1 + R_2$$

On the other hand, if these very same resistors are put in parallel, then their combined resistance is given by the different law:

$$1/R = 1/R_1 + 1/R_2$$

Order is clearly something which does play some role in the thinking of the physical scientist.

Actually, in fact, one cannot even claim that physics and chemistry can pay no attention to biological order. We have seen how biological genetics has had to move away from the idea that genes can be treated like beads randomly strung on a string. It is now realized that the order of the genes in a cell is of fundamental importance, and this order is obviously a case of biological order. However, it would be ridiculous to say that physics and chemistry, in particular molecular biology, is insensitive to this order. The very converse is the case. Molecular biology studies the order of the heritable material of the cell, right down to a single nucleotide pair. As we have seen, an absolutely key premise of molecular genetics is that although a DNA molecule contains only four kinds of base, a fantastic amount of information can be packed into the DNA molecule by virtue of the almost infinitely many ways in which these bases can be ordered along the molecule. Hence, if anything, this very same order is of more concern to the molecular biologist than it is to the non-molecular biologist.

Nevertheless, as in the case of the previous two arguments, there is an important grain of truth in the organismic argument about biological order. This is the fact that the scientist, particularly the physical scientist, must take biological order as given—he cannot hope to explain it away entirely given his present theories. For example, from the premises of his physico-chemical theories, the physical scientist might be able to explain why the DNA molecule is limited to the kinds of bases it has, and why the order of these bases has the effect it has. However, given the present theories, he does not seem able to explain why the order is as it is at the present. His situation seems similar to the situation of a physicist faced with current going through resistors. The physicist can, given his theories, explain the effect of the order as it is at the present; but he cannot explain how the order got to be the way that it is at the present. The reason why the resistors are in series rather than in parallel (or vice versa) lies outside of physics.

Could the molecular biologists ever hope to give a complete physico-chemical analysis of biological order? Even if he changes his theories will some order always have to be taken as given? My own feeling is that it will—any explanation of order at some particular time must incorporate information about order at some previous time (or, possibly, some future time). Of course, this is not to deny that as molecular biology develops, it will be possible to explain present biological order through a physico-chemical theory by referring to earlier and earlier ordered systems. Presumably, if one were to develop a full-blown physico-chemical evolutionary theory, theoretically one could start one's explanations by pre-supposing only order in the non-organic world. However, it seems fairly obvious that here is one point where theory and practice will never coincide. Hence, it does seem that there is merit in the organismic argument about biological order, although as I have just been explaining, I do not think this merit is quite as great as some of organismic biology's supporters think (see Schaffner, 1969b).

Before bringing this chapter to an end, one final comment must be made. I have just been arguing that I do not think the organismic arguments are as strong as some think they are. Nevertheless, even if what I claim is well-taken, I would not want it to be concluded that I am therefore arguing that the whole biological effort should henceforth be directed towards the elimination of biology as an independent discipline. The achievements of molecular biology are impressive and have implications for almost every area of biology. However, many areas of biology, for example areas like systematics and paleontology, are still essentially concerned with purely biological problems and can give answers only with the help of theories which are themselves biological. Possibly, if not almost certainly, in years to come we shall see the techniques and results of molecular biology playing a greater and greater role in solving problems posed by these areas like paleontology and systematics. But at the moment, our approach must be mainly biological, and hence to concentrate exclusively on molecular biology is to cease to attempt to solve problems in these areas. Such a cessation, it seems to me, implies as much a one-sided attitude as that of the traditional biologist who refuses to see any merit at all in the recent advances in molecular biology. (An interesting question, perhaps more one of psychology than philosophy, is why many of today's great biologists are adamantly opposed to any kind of biological reductionist thesis. Hein, 1969, has some interesting comments on this problem. I myself, Ruse 1971e, have tried to find some answers by using Kuhn's, 1962, notion of a 'paradigm'.)

POSTSCRIPT

I hope very much that this book will have convinced some philosophers that there are many areas of biology in which there lie interesting and important philosophical questions. I hope also the book will have convinced some biologists that the ideas and techniques of philosophy have a relevance to what they do as working scientists. I am very much conscious of the book's omissions, both philosophically and biologically. I am particularly aware of the fact that many areas of biology, for example, embryology, have been practically or entirely ignored. My silence about them should not be taken to indicate that I think them of no philosophical significance— I feel sure that, in fact, such areas harbour important problems awaiting discussion. Possibly the greatest omission in the book is any attempt to link up the biological sciences with the sciences about man. I am absolutely convinced that in the future, just as at one end biology is merging with the physical sciences, so at the other end biology will merge with the social sciences. More and more we shall see disciplines like psychology, sociology, and anthropology, incorporate into their theories results first discovered by the biologist. As this happens, I think the philosopher will have an increasingly important role to play, and I think that conversely, the meeting of the biological and social sciences may throw valuable light on such traditional philosophical problems as those of free-will and determinism, and the nature and relation of body and mind. Here, I have no room even to try to predict what will come from the confrontation of the biological and social sciences; but were I looking for another major programme in the philosophy of biology, it is in this meeting point between the two kinds of science that I would begin my search.

BIBLIOGRAPHY

ACHINSTEIN, P. (1968). *Concepts of Science*. Baltimore: Johns Hopkins Press.

ACHINSTEIN, P. (1971). *Law and Explanation: An Essay in the Philosophy of Science*. Oxford: Oxford University Press.

ALEXANDER, H. G. (1958). 'General statements as rules of inference', in *Minnesota Studies in the Philosophy of Science*, 2, H. Feigl, M. Scriven, and G. Maxwell (Eds.), 309–29.

AYALA, F. J. (1968). 'Biology as an autonomous science', *Am. Sci.* 56, 207–21.

BARKER, A. D. (1969). 'An approach to the theory of natural selection', *Philosophy*, XLIV, 271–90.

BECKNER, M. (1959). *The Biological Way of Thought*. New York: Columbia University Press.

BECKNER, M. (1967). 'Aspects of explanation in biological theory', in *Philosophy of Science Today*, S. Morgenbesser (Ed.). New York: Basic Books.

BECKNER, M. (1969). 'Function and teleology', *J. Hist. Bio.*, 2, 151–64.

BENZER, S. (1962). 'The fine structure of the gene', *Scientific American*, 206, 70–84.

BOWMAN, R. I. (1961). *Morphological Differentiation and Adaptation in the Galapagos Finches*. University of California Publications in Zoology, LVIII.

BRAITHWAITE, R. B. (1953). *Scientific Explanation*. Cambridge: Cambridge University Press.

BRIDGMAN, P. W. (1927). *The Logic of Modern Physics*. New York: Macmillan.

BRODBECK, M. (1962). 'Explanation, prediction, and "imperfect" knowledge', in *Minnesota Studies in the Philosophy of Science*, H. Feigl and G. Maxwell (Eds.). Minnesota: University of Minnesota Press, 231–72.

BUCK, R. C. and HULL, D. L. (1966). 'The logical structure of the Linnaean hierarchy', *Systematic Zoology*, 15, 97–111.

BUCK, R. C., and HULL, D. L. (1969). 'A reply to Gregg', *Systematic Zoology*, 18, 354–7.

CAIN, A. J. (1954). *Animal Species and their Evolution*. London: Hutchinson.

CANNON, H. G. (1955). 'What Lamarck really said', *Proc. Lin. Soc. London*, 168, 70–87.

CANNON, H. G. (1958). *Lamarck and Modern Genetics*. Manchester: Manchester University Press.

CARLO, W. E. (1967). *Philosophy, Science and Knowledge*. Milwaukee: Bruce.

CARLSON, E. A. (1966). *The Gene: A Critical History*. Philadelphia: Saunders.

CARTER, G. S. (1951). *Animal Evolution*. London: Sidgwick and Jackson.

CAUSEY, R. (1969). 'Polanyi on structure and reduction', *Synthese*, 20, 230–7.

CSONKA, P. L. (1969). 'Advanced effects in particle physics', I. *Physical Review*, **180**, 1266–81.

DARWIN, C. (1959 ed.). *The Origin of Species by Means of Natural Selection.* 1st ed. 1859, Variorum text, M. Peckham (Ed.), 1959. Philadelphia: University of Pennsylvania Press.

DAVIS, P. H., and HEYWOOD, V. H. (1963). *Principles of Angiosperm Taxonomy.* Edinburgh: Oliver and Boyd.

DELBRÜCK, M. (1949). 'A physicist looks at biology', *Transactions of the Connecticut Academy of Arts and Sciences*, **38**, 173–90.

DELEVORYAS, T. (1964). 'The role of palaeobotany in vascular plant classification', in *Phenetic and Phylogenetic Classification*, V. H. Heywood and J. McNeill (Eds.). London: The Systematics Association. 29–36.

DOBZHANSKY, TH. (1937). *Genetics and the Origin of Species.* 3rd ed. 1951. New York: Columbia University Press.

DOBZHANSKY, TH. (1956). 'What is an adaptive trait?' *American Naturalist*, **XC**, 337–47.

DOBZHANSKY, TH. (1962). *Mankind Evolving.* New York: Columbia University Press.

DOBZHANSKY, TH. (1970). *Genetics of the Evolutionary Process.* New York: Columbia University Press.

EHRLICH, P. R. (1964). 'Some axioms of taxonomy', *Systematic Zoology*, **13**, 109–23.

EHRLICH, P. R., and HOLM, R. W. (1962). 'Patterns and population', *Science*, **137**, 652–7.

EHRLICH, P. R., and RAVEN, P. H. (1969). 'Differentiation of populations', *Science*, **165**, 1228–32.

FALCONER, D. S. (1961). *Introduction to Quantitative Genetics.* New York: Ronald Press.

FEYERABEND, P. K. (1962). 'Explanation, reduction, and empiricism', in *Minnesota Studies in the Philosophy of Science*, **3**, H. Feigl and G. Maxwell (Eds.). Minneapolis: University of Minnesota Press, 28–97.

FISHER, R. A. (1936). 'Has Mendel's work been rediscovered?' *Annals of Science*, **1**, 115–37. Reprinted in *The Origins of Genetics: A Mendel Source Book*, C. Stern and E. R. Sherwood (Eds.). San Francisco: Freeman. 1966, 139–72.

FORD, E. B. (1964). *Ecological Genetics.* London: Methuen.

GALLIE, W. B. (1955). 'Explanations in history and the genetic sciences', *Mind*, **64**, 160–80.

GEORGE, W. (1964). *Elementary Genetics.* London: Macmillan.

GHISELIN, M. (1966). 'On psychologism in the logic of taxonomic controversies', *Systematic Zoology*, **15**, 207–15.

GIBSON, Q. (1960). *The Logic of Social Enquiry.* London: Routledge and Kegan Paul.

GLASS B. (1963). 'The relation of the physical sciences to biology—indeterminacy and causality', in *Philosophy of Science: The Delaware Seminar*, **1**, B. Baumrin (Ed.). New York: Interscience. 223–49.

GOLDSCHMIDT, R. B. (1940). *The Material Basis of Evolution.* New Haven: Yale University Press.

GOLDSCHMIDT, R. B. (1952). 'Evolution as viewed by one geneticist', *American Scientist*, **40**, 84–135.

GOUDGE, T. A. (1961). *The Ascent of Life.* Toronto: University of Toronto Press.

GREGG, J. R. (1954). *The Language of Taxonomy.* New York: Columbia University Press.

GREGG, J. R. (1967). 'Finite Linnaean structures', *Bulletin of Mathematical Biophysics*, **29**, 191–206.

GREGG, J. R. (1968). 'Buck and Hull: a critical rejoinder', *Systematic Zoology*, **17**, 342–4.

GRUNER, R. (1966). 'Teleological and functional explanations', *Mind*, **LXXV**, 516–26.

GRUNER, R. (1969). 'Uniqueness in nature and history', *Phil. Quart.*, **19**, 145–54.

HALDANE, J. B. S. (1964). 'A defense of bean bag genetics', *Pers. Biol. Med.*, **7**, 343–59.

HANSON, N. R. (1958). *Patterns of Discovery*. Cambridge: Cambridge University Press.

HEIN, H. (1969). 'Molecular biology vs organicism', *Synthese*, **20**, 238–54.

HELMER, O., and RESCHER, N. (1959). 'On the epistemology of the inexact sciences', *Management Science*, **6**, 25–52.

HEMPEL, C. G. (1954). 'A logical appraisal of operationism', *Scientific Monthly*, **79**, 215–20. Reprinted in C. G. Hempel, *Aspects of Scientific Explanation*, 1965, New York: The Free Press, 123–33.

HEMPEL, C. G. (1959). 'The logic of functional analysis', in *Symposium on Sociological Theory*, L. Gross (Ed.). New York: Harper and Row, 271–307.

HEMPEL, C. G. (1965). 'Aspects of scientific explanation', in *Aspects of Scientific Explanation and Other Essays in the Philosophy of Science*. New York: The Free Press.

HEMPEL, C. G. (1966). *Philosophy of Natural Science*. Englewood, N.J.: Prentice-Hall.

HEMPEL, C. G., and OPPENHEIM, P. (1948). 'Studies in the logic of explanation', *Phil. Sci.*, **15**, 135–75.

HIMMELFARB, G. (1962). *Darwin and the Darwinian Revolution*. New York: Anchor.

HULL, D. L. (1967). 'Certainty and circularity in evolutionary taxonomy', *Evolution*, **21**, 174–89.

HULL, D. L. (1968). 'The operational imperative: Sense and nonsense in operationism', *Systematic Zoology*, **17**, 438–57.

HULL, D. L. (1970a). 'Contemporary systematic philosophies', *Annual Review of Ecology and Systematics*, **1**, 19–54.

HULL, D. L. (1970b). 'Morphospecies and biospecies: a reply to Ruse', *Brit. J. Phil. Sci.*, **21**, 280–2.

JEPSON, G. L. (1949). 'Selection, "orthogenesis", and the fossil record', *Proc. Am. Phil. Soc.*, **93**, 479–500.

KETTLEWELL, H. B. D. (1955). 'Selection experiments on industrial melanism in the *Lepidoptera*', *Heredity*, **9**, 323–42.

KÖRNER, S. (1966). *Experience and Theory: An Essay in the Philosophy of Science*. London: Routledge and Kegan Paul.

KUHN, T. S. (1962). *The Structure of Scientific Revolutions*. Chicago: University of Chicago Press.

LACK, D. (1947). *Darwin's Finches: An Essay on the General Biological Theory of Evolution*. Cambridge: Cambridge University Press.

LACK, D. (1954). *The Natural Regulation of Animal Numbers*. Oxford: Oxford University Press.

LACK, D. (1966). *Population Studies of Birds*. Oxford: Oxford University Press.

LAMARCK, J. B. (1809). *Zoological Philosophy*. Trans. H. Elliot. 1963. New York: Hafner.

LEHMAN, H. (1965a). 'Functional explanation in biology', *Phil. Sci.*, **32**, 1–20.

LEHMAN, H. (1965b). 'Teleological explanation in biology', *Brit. J. Phil. Sci.*, **15**, 327.

LEHMAN, H. (1967). 'Are biological species real?' *Phil Sci.*, **34**, 157–67.
LEWONTIN, R. C. (1970). 'The units of selection', *Annual Review of Ecology and Systematics*, **1**, R. F. Johnston, *et al.* (Eds.). California: Annual Reviews Inc., 1–18
LI, C. C. (1955). *Population Genetics*. Chicago: Chicago University Press.
LIVINGSTONE, F. B. (1967). *Abnormal Hemoglobins in Human Populations*. Chicago: Aldine.
LIVINGSTONE, F. B. (1971). 'Malaria and human polymorphisms', *Annual Review of Genetics*, **5**, 33–64.
MACKIE, J. L. (1966). 'The direction of causation', *Phil. Review*, **LXXV**, 441–66.
MANSER, A. R. (1965). 'The concept of evolution', *Philosophy*, **XL**, 18–34.
MAYR, E. (1942). *Systematics and the Origin of Species*. New York: Columbia University Press.
MAYR, E. (Ed.) (1957). *The Species Problem*. Washington, D.C.: A.A.A.S. Pub. 50.
MAYR, E. (1963). *Animal Species and Evolution*. Cambridge, Mass.: Belknap.
MAYR, E. (1965). 'Numerical phenetics and taxonomic theory', *Systematic Zoology*, **14**, 73–97.
MAYR, E. (1969). *Principles of Systematic Zoology*. New York: McGraw-Hill.
MAYR, E. (1972). 'Lamarck revisited', *J. Hist. Bio.*, **5**, 55–94.
MICHENER, C. D. (1963). 'Some future developments in taxonomy', *Systematic Zoology*, **13**, 151–72.
MONTEFIORE, A. (1956). 'Professor Gallie on "necessary and sufficient conditions" ', *Mind*, **65**, 534–41.
MULLER, H. J. (1939). 'Reversibility in evolution considered from the standpoint of genetics', *Biol. Rev.*, **14**, 261–80.
MUNSON, R. (1971). 'Biological adaptation', *Phil. Sci.*, **38**, 200–215.
MUNSON, R. (1972). 'Biological adaptation: a reply', *Phil. Sci.*, **39**.
NAGEL, E. (1961). *The Structure of Science*. London: Routledge and Kegan Paul.
POLANYI, M. (1968). 'Life's irreducible structure', *Science*, **160**, 1308–12.
RACE, R. R., and SANGER, R. (1954). *Blood Groups in Man*. 2nd ed. Oxford: Blackwell.
RAPER, A. B. (1960). 'Sickling and malaria', *Trans. Roy. Soc. Trop. Med. Hyg.*, **54**, 503–4.
RENSCH, B. (1960). 'The laws of evolution', in *Evolution after Darwin*, Vol. I, Sol Tax. Chicago: Chicago University Press, 95–116.
ROMER, A. S. (1941). *Man and the Vertebrates*. Chicago: Chicago University Press.
ROSS, H. H. (1964). 'Review of *Principles of Numerical Taxonomy*', *Systematic Zoology*, **13**, 108.
RUDWICK, M. J. S. (1964). 'The inference of function from structure in fossils', *Brit. J. Phil. Sci.*, **15**, 27–40.
RUDWICK, M. J. S. (1972). *The Meaning of Fossils*. London: Macdonald.
RUSE, M. (1969). 'Definitions of species in biology', *Brit. J. Phil. Sci.*, **20**, 97–119.
RUSE, M. (1970a). 'The revolution in biology', *Theoria*, **XXXVI**, 1–22.
RUSE, M. (1970b). 'Are there laws in biology?', *Aust. J. Phil.*, **48**, 234–46.
RUSE, M. (1971a). 'Natural selection in *The Origin of Species*', *Studies in History and Philosophy of Science*, **1**, 311–51.
RUSE, M. (1971b). 'The species problem: a reply to Hull', *Brit. J. Phil. Sci.*, **22**, 369–71.
RUSE, M. (1971c). 'Gregg's Paradox: A proposed revision to Buck and Hull's solution', *Systematic Zoology*, **20**, 239–45.
RUSE, M. (1971d). 'Reduction, replacement, and molecular biology', *Dialectica*, **25**, 39–72.

RUSE, M. (1971e). 'Two biological revolutions', *Dialectica*, **25**, 17–38.

RUSE, M. (1971). 'Functional statements in biology', *Phil. Sci.*, **38**, 87–95.

RUSE, M. (1972). 'Biological adaptation', *Phil. Sci.*, **39**.

RYLE, G. (1950). '"If", "so", and "because",' in *Philosophical Analysis*, M. Black (ed.), Englewood Cliffs: Prentice-Hall.

SCHAFFNER, K. F. (1967a). 'Approaches to reduction,' *Phil. Sci.*, **34**, 137–47.

SCHAFFNER, K. F. (1967b). 'Antireductionism and molecular biology', *Science*, **157**, 644–7.

SCHAFFNER, K. F. (1969a). 'The Watson–Crick model and reductionism', *Brit. J. Phil. Sci.*, **20**, 325–48.

SCHAFFNER, K. (1969b). 'Theories and explanations in biology', *J. Hist. Bio.*, **2**, 19–33.

SCHAFFNER, K. F. (1969c). 'Chemical systems and chemical evolution. The philosophy of molecular biology', *American Scientist*, **57**, 410–20.

SCHINDEWOLF, O. H. (1950). *Grundfragen der Paläontologie*. Stuttgart: Schweizerbart.

SCHLESINGER, G. (1963). *Method in the Physical Sciences*. New York: The Humanities Press.

SCRIVEN, M. (1959). 'Explanation and prediction in evolutionary theory', *Science*, **130**, 477–82.

SCRIVEN, M. (1961). 'The key property of physical laws—inaccuracy', in *Current Issues in the Philosophy of Science*, H. Feigl and G. Maxwell (Eds.). New York: Holt, Reinhart and Winston, 91–101.

SIMPSON, G. G. (1951). 'The species concept', *Evolution*, **5**, 285–98.

SIMPSON, G. G. (1953). *The Major Features of Evolution*. New York: Columbia University Press.

SIMPSON, G. G. (1963a). 'Historical science', in *The Fabric of Geology*, C. C. Albritton Jr. (Ed.). Stanford, California: Freeman, Cooper and Co. 24–48.

SIMPSON, G. G. (1963b). *This View of Life*. New York: Harcourt, Brace and World.

SKLAR, A. (1964). 'On category overlapping in taxonomy', in *Form and Strategy in Science*, J. R. Gregg and F. T. C. Harris (Eds.). Dordrecht: Reidel.

SMART, J. J. C. (1963). *Philosophy and Scientific Realism*. London: Routledge and Kegan Paul.

SOKAL, R. R., and ROHLF, F. J. (1970). 'The intelligent ignoramus, an experiment in numerical taxonomy', *Taxon.*, **19**, 305–19.

SOKAL, R. R., and SNEATH, P. H. A. (1963). *Principles of Numerical Taxonomy*. San Francisco: W. H. Freeman and Co.

SOMMERHOFF, G. (1950). *Analytical Biology*. Oxford: Oxford University Press.

SPECTOR, M. (1966). 'Theory and observation', *Brit. J. Phil. Sci.*, **XVII**, 1–20, 89–104.

STEBBINS, G. L. (1950). *Variation and Evolution in Plants*. New York: Columbia University Press.

STURTEVANT, A. H. (1925). 'The effects of unequal crossing over at the bar locus in *Drosophila*', *Genetics*, **10**, 117–47.

TAYLOR, C. (1964). *The Explanation of Behaviour*. London: Routledge and Kegan Paul.

THODAY, J. M., and GIBSON, J. B. (1962). 'Isolation by disruptive selection', *Nature*, **193**, 1164–6.

TOULMIN, S. (1961). *Foresight and Understanding*. London: Hutchinson.

VAN VALEN, L. (1964). 'An analysis of some taxonomic concepts', in *Form and Strategy in Science*, J. R. Gregg and F. T. C. Harris (Eds.). Dordrecht: Reidel.

Von Bertalanffy, L. (1952). *Problems of Life*. New York: John Wiley.

Waddington, C. H. (1957). *The Strategy of the Genes*. London: George Allen and Unwin.

Waddington, C. H. (Ed.) (1969). *Towards a Theoretical Biology*. 2, *Sketches*. Edinburgh: Edinburgh University Press.

Watson, J. D. (1965, 2nd ed. 1970). *Molecular Biology of the Gene*. New York: Benjamin.

Whewell, W. (1840). *Philosophy of the Inductive Sciences*. London: Parker.

Whitehouse, H. L. K. (1965). *Towards an Understanding of the Mechanism of Heredity*. New York: St. Martin's Press.

Williams, G. C. (1966). *Adaptation and Natural Selection: A Critique of Some Current Evolutionary Thought*. Princeton: Princeton University Press.

Williams, M. B. (1970). 'Deducing the consequences of evolution: a mathematical model', *J. Theoretical Biology*, **29**, 343–85.

Wimsatt, W. C. (1970). 'Some problems with the concept of "feedback" ', *Boston Studies in the Philosophy of Science*, **8**, 241–56.

Wimsatt, W. (1972). 'Teleology and the logical structure of function statements', *Stud. Hist. Phil. Sci.*, **3**, 1–80.

Woodger, J. H. (1952). *Biology and Language*. Cambridge: Cambridge University Press.

Wright, L. (1972). 'A comment on Ruse's analysis of function statements', *Phil. Sci.*, **39**.

Wright, S. (1931). 'Evolution in Mendelian populations', *Genetics*, **16**, 97–159.

Wright, S. (1966). 'Mendel's Ratios', in *The Origins of Genetics: A Mendel Source Book*, C. Stern and E. R. Sherwood (Eds.). San Francisco: W. H. Freeman. 173–5.

Wynne-Edwards, V. C. (1962). *Animal Dispersion in relation to Social Behaviour*. Edinburgh: Oliver and Boyd.

Young, R. M. (1971). 'Darwin's metaphor: does nature select?' *Monist*, **55**, 442–503.

INDEX